Brian Harvey

The Rebirth of the Russian Space Program

50 Years After Sputnik, New Frontiers

 Springer

Published in association with
Praxis Publishing
Chichester, UK

Mr Brian Harvey
Terenure
Dublin 6W
Ireland

SPRINGER–PRAXIS BOOKS IN SPACE EXPLORATION
SUBJECT *ADVISORY EDITOR*: John Mason, M.Sc., B.Sc., Ph.D.

ISBN 978-0-387-71354-0 Springer Berlin Heidelberg New York

Springer is part of Springer-Science + Business Media (springer.com)

Library of Congress Control Number: 2007922812

Cover design: Jim Wilkie
Project management: Originator Publishing Services Ltd, Gt Yarmouth, Norfolk, UK

Printed on acid-free paper

Contents

Author's preface

The rebirth of the Russian space program marks an important event: 50 years since the first Sputnik was launched on 4th October 1957. At that time, few could have imagined the dramatic events that lay head. The Soviet Union achieved all the great firsts in cosmonautics—the first satellite in orbit, the first animal in orbit, the first laboratory in orbit, the first probe to the Moon, the first probe to photograph its far side, the first soft landing on the moon, the first man in space, the first woman in space, the first spacewalk. Except one, the first human landing on the Moon. In 1964, the Soviet Union decided to contest the decision of the United States to put the first person on the Moon. The Soviet Union engaged in that race far too late, with divided organization, and made a gallant but doomed challenge to Apollo.

Undaunted, the Soviet Union rebuilt its space program around orbiting stations, building the first one, Salyut, and then the first permanent home in space, Mir. The Soviet Union still achieved many more firsts: the first lunar rover, the first soft landing on Venus, the first soft landing on Mars, the first recovery of samples from the Moon by automatic spacecraft.

The original book in this series *Race into space—a history of the Soviet space programme* (1988) was written during the heyday of the Soviet space program, when the Soviet Union was launching over a hundred satellites a year and had a vast program for the manned and unmanned exploration of space and its application for practical benefits on Earth. The second book in the series, *The new Russian space program* (1995) was compiled during the shock adjustment of the former Soviet space program to the strained economic realities of life in the Russian Federation. *The new Russian space program* took advantage of all the new information that had come to light about Soviet history: thanks to the policy of openness (*glasnost*) begun by the last president of the Soviet Union, Mikhail Gorbachev, we were now at last able to learn about what had really taken place during the time of the Soviet period, as secrets emerged into the light of day. The third book of this series, *Russia in space—the failed frontier?* (2001) looked at the Russian space program in the period from 1992, when the Russian Federation came into existence. The title of course posed a question: Had the Russian romance with cosmonautics run its course and "failed"? This book chronicled the decline and difficulties of the 1990s, but also showed how the program had adapted, survived and, sometimes grimly, held on.

Now, the fourth book in the series, *The rebirth of the Russian space program* looks at the Russian space program at a convenient marking point, fifty years after Sputnik. It chronicles developments since the turn of the century, takes a look at the Russian program as it is now and looks toward the future. This account focuses on the years 2000 to 2006. Readers who wish to study the earlier history should return to *Race into space* (for the Soviet period), *The new Russian space programme* (for the transition) and *Russia in space—the failed frontier?* (for the 1990s). By way of definitions, this book covers primarily the Russian Federation—but it does take in those parts of the Ukrainian space program rooted in the old Soviet program.

The rebirth of the Russian space program coincides not only with Sputnik but with the announcement by the government of the federation of a space plan to last to 2015, an attempt to reinstitute goal-orientated planning in the program. If one looks at the number of launches per year, Russia remains the leading spacefaring nation in the world. At the same time, it is obvious that the Russian Federation's space program has none of the ambition of the American space program, which has now sent extraordinary missions to all the corners of the solar system and plans to return astronauts to the Moon and send them onward to Mars. Unlike the 1960s and 1980s, present-day Russia has neither the capacity nor the will to challenge American leadership of space exploration (the Chinese probably do, but that is another story). At the same time, Russia will remain one of the world's space superpowers, a builder of space stations, a formidable contributor to the world space industry and science. As this book shows, the Russian space program is full of activity and life. Fifty years after Sputnik, the dream lives on.

Brian Harvey
Dublin, Ireland, 2007

Acknowledgments

The author wishes to thank all those who contributed to this book through the provision of ideas, information, comments, illustrations and photographs. In particular, he would like to thank: Rex Hall, Phil Clark, Bart Hendrickx and of course Clive Horwood, the publisher, for his support and encouragement.

Photographs in the book come from the author's collection and from illustrations published in the previous three editions of the book. He renews his thanks to those who contributed. For this edition, he wishes to especially thank:

- V.K. Chvanov in Energomash, for illustrations of rockets and engines;
- Roman Turkenich in NPO PM Krasnoyarsk, for illustrations of the applications satellites built by NPO PM;
- V.M. Vershigorov in TsSKB Progress for images of the R-7 rocket and variants, Resurs and the work of TsSKB;
- Bert Vis, for his photographs of the US KMO;
- Nicolas Pillet, for his images of Kliper;
- Dominic Plelan, for his photograph of the *botchka*;
- Peter Freeborn, Eurockot, for images of Rockot and Plesetsk Cosmodrome; and
- NASA for images of the missions to the International Space Station.

Brian Harvey
Dublin, Ireland, 2007

About the book

First published as *Race into space—a history of the Soviet space programme*, Ellis Horwood, 1988

Re-published as *From competition to collaboration—the new Russian space programme*, 2nd edition, by Wiley/Praxis, 1996

Re-published as *Russia in space—the failed frontier?* 3rd edition, published by Springer/Praxis, 2001

Figures

Maps

Tables

Abbreviations and acronyms

3SL	Three-stage Sea Launch (Zenit)
CADB	Chemical Automatics Design Bureau
CIS	Commonwealth of Independent States
CNES	Centre National d'Études Spatiales
comsat	communications satellite
COROT	COnvection, ROtation and planetary Transits
COSPAR	COmmittee on SPace Research
COSPAS-SARSAT	Search and Rescue Satellite System
CSTS	Crew Space Transportation System
DELTA	Dutch Expedition for Life, Technology and Atmospheric (research program)
DK	Dmitri Kozlov (satellite)
DMC	Disaster Monitoring Constellation
DMS-R	Data Management System
DS	Dnepropetrovsky Sputnik
DSP	Defence Support Program
elint	electronic intelligence
ELS	Ensemble de Lancement Soyuz
EOB	Electronic Order of Battle
EORSAT	Elint Ocean Reconnaissance SATellite
ESA	European Space Agency
EVA	Extra Vehicular Activity
FG	Forsunochnaya Golovka (Fuel Injector)
FGB	*Funkstionalii Gruzovoi Blok* (Functional Cargo Block)
FLPP	Future Launcher Preparatory Program
FLTP	Future Launchers Technology Program
GDL	Gas Dynamics Laboratory
GFZ	Geoforschungszeutrum (Geological Research Centre)

GLONASS	*Globalnaya Navigatsionnaya Sputnikovaya Sistema*
GPS	Global Positioning System
GRACE	Gravity Recovery And Climate Experiment
GRU	Glavnoye Razvedyvatelnoye Upravleniye (main military intelligence directorate)
GSLV	GeoSynchronous Launch Vehicle
HIV	Human Immunodeficiency Virus
IAF	International Astronautical Federation
ICM	Interim Control Module
IKI	Institute for Space Research
ILS	International Launch Services
IM	*Issledovatl Modul*
IMBP	Institute for Medical and Biological Problems
INDEX	INnovative technology Demonstration EXperimental (satellite)
INMARSAT	International Mobile Satellite Organization
IP	Instrument Point
IRDT	Inflatable Reentry and Descent Technology
ISS	International Space Station
JPL	Jet Propulsion Laboratory
KB	Design bureau
KBKhA	*KB KhimAutomatiki* (Chemical Automatics Design Bureau)
KH	Key Hole (code name for American reconnaissance satellites)
KOMPASS	Complex orbital magneto plasma autonomous small satellite
Koronas	Comprehensive Orbital Near Earth Observations of the Active Sun
KTOK	Ccomplex for simulators for spaceships
MARSPOST	MARS Piloted Orbital STation
metsat	meteorological satellite
MIK	Integration and test building hall
MOM	*Ministerstvo Obshchego Machinostroyeniye* (Ministry of General Machine Building)
MPLM	Multi-Purpose Laboratory Module
MSNBC	Microsoft and NBC Universal News
NASA	National Aeronautics and Space Administration
NEM	*Nauk Energiya Modul*
NII	*Nauk Issledovatl Institut* (scientific research institute)
NIP	Scientific instrument points
NITsPlaneta	Scientific Research Center of Space Meteorology
NPO	Scientific and production association
NSAU	National Space Agency of Ukraine
OICETS	Optical Interorbit Communications Engineering Test Satellite

OKB	*Opytnoye Konstruktorskoye Buro* (experimental design bureau)
OSETS	*Orbitalny Sborochno Eksploratsionny Tsentr* (Orbital Assembly and Operations Centre)
PAMELA	Payload for Anti Matter Exploration and Light nuclei Astrophysics
RAN	Russian Academy of Sciences
RD	*Raketny Dvigatel* (rocket motor)
RGRL	Roentgen Gamma Rosita Lobster
RKA	Russian Space Agency
RKK	Rocket Cosmic Corporation
SKKP	System for Monitoring Space
SpKs	*Spuskayemaya Kapsula*
SRN	State security
SSTL	Surrey Satellite Technology Ltd.
START	STrategic Arms Reduction Talks
T	Transport (Soyuz version)
TDRS	Tracking and Data Relay System
TM	Transport Modified (Soyuz version)
TMA	*Transport Modified Anthropometric* (Soyuz version)
TMK	Tizhuly Mezhplanetny Korabl (Heavy Interplanetary Spaceship)
TORU	*Tele Operatorny Rezhim Upravleniye* (Television Remote Control)
TsAGI	Central Institute for Aero Hydrodynamics
TsDUC	Center for Long-Range Space Communications
TsPK	Yuri Gagarin Cosmonaut Training Center
TsSKB	*Tsentralnoye Spetsializorovannoye Konstruktorskoye Buro* (Central Specialized Design Bureau)
TsUP	*Tsentr Upravleniye Polyotami*
TUB	Technical University of Berlin
UDM	Universal Docking Module
UDMH	Unsymetrical Dimethyl Methyl Hydrazine
US A	*Upravleniye Sputnik* (controlled sputnik, "active")
US P	*Upravleniye Sputnik* (controlled sputnik, "passive')
VDNK	Exhibition of Economic & Scientific Achievements
VEGA	Venus Halley
VHF	Very High Frequency
VMK	Recoverable Maneuverable Capsules
VPK	Commission on Military Industrial Issues 280

1

Almost the end

The 25th June 1997 was a really, really bad day in the Russian space program. It is not too much to say that it all nearly came to an end that one day.

On board the Mir space station, the flagship project of the Soviet and Russian space fleet, it started as a routine day. Mir was in its twelfth year circling the Earth. Crewing the station were two cosmonauts, Vasili Tsibliev and Alexander Lazutkin, as well as an American visitor, British-born Michael Foale. Tsibliev and Lazutkin had come on board in March, Foale the previous month. They had made Mir their home and there they lived, ate, slept, watched the Earth, exercised to combat weightlessness and carried out the many experiments for which the orbital station had been built. Mir was a huge complex, more than 100 tonnes in weight, comprising a base block (Mir) modules (Kvant, Kvant 2, Krystall, Spektr, Priroda), manned spacecraft (Soyuz), unmanned freighters called Progress and beams and girders. There was a central node where modules docked, control panels, laboratories, gardens, exercise machines and sleeping berths. Air and electricity cables snaked around the walls and through the tunnels. Mir was made homely by posters on the walls, a video library, even a bookshelf.

The principal task of the day for Vasili Tsibliev was to guide in to Mir's docking port a robot spacecraft, Progress M-34, using a remote controller called TORU. Normally, unmanned Progress spacecraft came in using an automated Ukrainian system called Kurs, but since the break-up of the Soviet Union the Ukrainians charged a huge amount for Kurs docking systems, so the Russians were trying something cheaper. The TORU was a small joystick, very much like a controller used by a child on a computer game or play station. In front of him was a television picture, beaming him the image of Mir from Progress as it closed in, along with display data and grids superimposed on the screen. Using the TORU controller, he would send radio commands to Progress to fire its thrusters to move faster, slower, up or down, left or right. TORU was a simple, effective system, costing a hundredth the price of the old Kurs.

The TORU had not been used before and Vasili Tsibliev, the space station commander, was apprehensive, but it was no more challenging than hundreds of similar tasks that he had undertaken while a cosmonaut. That was what he was trained for. The previous day he had commanded the seven-tonne Progress M-34 freighter to separate from the orbital station and let it drift away for a couple of kilometers. Now, he sent the signal for it to come back, guided in by the TORU.

What seems to have happened was this. Tsibliev commanded the freighter to come in. He asked his colleague Alexander Lazutkin to position himself near one of the windows to help him spot the arriving Progress. On the television set, it was extraordinarily difficult for the Progress television to pick out Mir from a distance against the speckled clouds of Earth underneath. Progress seemed to come in too slowly. So, Tsibliev fired its thrusters to make it come in faster. Tsibliev and Lazutkin used a set of squares on the screen to measure the distance (a square on the grid meant it was 5 km out). Using a combination of the grid, a stopwatch and the television camera, Tsibliev steered Progress in. When it was 1 km out, or so he thought, Tsibliev applied a standard braking maneuver, in order to slow Progress to walking pace for the final approach. All the time, he had Lazutkin watching out. On the screen, Mir was much bigger now, filling four grids on the square. This time, Tsibliev was alarmed at its rapid rate of approach. He fired the thrusters repeatedly to slow the freighter, but to no avail. He and Lazutkin dashed from one window to the other, trying to spot Progress and they enlisted the visiting astronaut Michael Foale in the effort. According-ing to the plan, Progress should now have been 400 m out.

But it was too late. Lazutkin at last spotted the Progress, not at the 400 m on the worksheet but at 150 m and closing rapidly. "It's here already!" yelled Lazutkin.

Vasili Tsibliev

Michael Foale

Progress careered into one of the laboratory modules at some speed, crumpled its solar panels and drifted off to the side. Next thing, the astronauts felt a pop in their ears: the pressure in the station had begun to drop. The station had been hit and was punctured. Air was hissing out of the space station. The master alarm at once rang out. The space station's manual stated that a pressure loss would empty air out of the station in 18 min: they would have to either evacuate the station within that time, or seal the leak or they would be dead. Tsibliev yelled to Foale in Russian: "Va korabl!" Foale understood Russian, but in English this meant: get into the attached Soyuz spaceship to return to Earth!

Foale pushed himself quickly down the tunnel, into the node, into Soyuz at the far end. Once there, he removed the hoses and cables going from the node into Soyuz, so as to prepare it for emergency descent. Once he got there, he realized that his colleagues had no immediate plans to join him in a fast plunge back to Earth: they were trying to save the station first. You did not just abandon the pride of the Russian space fleet without a fight.

Michael Foale was soon to realize that it was the Spektr laboratory that had taken the hit. Spektr was, after all, his module, where he slept and carried out his experiments. In theory, it was simple enough to close the hatch between Spektr and the node, but the hatchway was full of cables and ducts. He at once rejoined his colleagues. Lazutkin and Foale used a knife to cut the cables, sparks flying and tried to close the hatch between the node and the module. It would not close—because the escaping air was pulling it outward. However, the two men found a hatch cover on their side which they jammed in its place from the node side, the escaping air sealing it in. The leak was now on the other side of the hatch and they were saved. The job took 14 min and the air stopped venting at once. Pressure had fallen from 760 mm to 693 mm.

Disaster had been averted, but a whole series of knock-on problems had only just begun. Cutting the cables from Spektr meant that the main part of Mir lost all the solar electric power coming from Spektr. This was considerable, for Spektr had four large panels which supplied 40% of Mir's energy requirements. Tsibliev, Lazutkin and Foale powered down Mir's equipment, abandoned scientific work for the time being and learned to live by torchlight. Experiments and non-essential equipment were powered down.

Mission control was informed on a crackly line during the next pass over Russian territory. Gradually, the three men tried to restore the situation and within two days they had got pressure on the station back up to 770 mm. But, just as they were beginning to return to normal, the station drifted out of alignment from the Sun and lost all its remaining solar power on 3rd July. Its gyrodynes powered down. Though stable, Mir was now drifting helplessly in Earth orbit, unable to lock on to the Sun and acquire its electricity-giving powers. The ventilation system was silent. The normal clatter and hum of the orbiting station was replaced by dead calm. With vents turned off, Mir became a silent station, eerily so, like a ship drifting in the horse latitudes of old.

So they climbed in the small Soyuz cabin, the spaceship that would bring them back to Earth, and used its scarce fuel to turn the whole complex toward the Sun. This did the trick and Mir's panels were able to acquire sufficient sunlight to get a flicker of power back into the system. They locked on. It took a full day to power the station up again. Foale had used up 70 kg of Soyuz' fuel reserves in doing so.

But then on 16th July, Alexander Lazutkin, in the course of normal operations controlling the station, accidentally pulled out the main computer cable in the guidance system. This sparked another master alarm and, worse, shut the system's navigation system down, causing Mir to drift out of alignment and lose power. This time the cosmonauts knew what to do and got Mir back on line again within a day. Once again they had to retreat to the Soyuz, explain the situation to ground control on Soyuz' radio and use the Soyuz thrusters.

Alexander Lazutkin

Mir

Two days later, Vasili Tsibliev was to suffer a personal setback. As he ran his daily exercises, the doctors at ground control detected an irregularity in his heartbeat. Quite simply, the strain of the events of the previous weeks had got to him. Plans to do an internal spacewalk to fix the leak on Spektr were shelved, for this crew anyway.

At this stage, things on Mir could not get much worse. Round the world, the media had followed events on the station, running apocalyptical headings about the disaster-prone space station. Some Americans wanted their astronauts withdrawn from the Mir program: it was just too dangerous.

THE ADMIRATION OF THE WORLD

During the Mir crisis, it was hard to avoid the conclusion that the Russian space program was now close to collapse. But, ten years earlier, in 1987, the Soviet space program had been the admiration of the world. Even the Americans had been forced to admire and one had no further to look than some of the great barometers of American opinion. The *National Geographical Magazine* had just run an issue *The Soviets in space* (October 1986). *Time* magazine ran a self-explanatory headline *Surging ahead* (5th October 1987).

In 1987, about 400,000 people worked for the Soviet space program—in the assembly lines in Moscow, in rocket factories in Dnepropetrovsk and Kyubyshev, in scientific institutes scattered around the country, in production plants and in the worldwide land and sea-based tracking network. The three cosmodromes of Baikonour, Plesetsk and Kapustin Yar launched a hundred satellites a year (102 in one year), an average of two a week. There were military satellites for photo-reconnaissance, electronic intelligence and even for spies abroad to send their messages home. In the area of applications, Soviet satellites circled the Earth for communications (Molniya), weather-forecasting (Meteor) and television (Raduga, Gorizont, Ekran). In science, missions were flown with monkeys and other animals (Bion) and to carry out materials-processing in zero gravity (Foton).

The last Soviet deep space mission had been stunning. Two spaceships had been sent to Venus, where probes had already landed on the planet, drilling and analyzing its rocks, while other probes had made radar maps. Now VEGA 1 and 2 were launched there. When they arrived, they dropped landers down to the surface, one to a lowland area, the Mermaid Plains, the other to highlands, Aphrodite, to measure the atmosphere and surface. VEGA 1 and 2 dropped balloons into the atmosphere to travel a bumpy two days in its acid clouds. Meantime, the mother ships altered course across the solar system, intercepting Comet Halley and taking amazing blue, red and white pictures of its close encounter relayed live to a mission control full of foreign scientists.

1987 saw the launch of the most powerful rocket in the world, the Energiya, the ultimate creation of the chief designer of the Soviet space program, Valentin Glushko. He had been building rocket engines in Leningrad since the 1920s and in May 1987 his Energiya made its first flight. The new Energiya—a name meaning "energy" in Russian—took 12 sec to build up to full thrust. Weighing 2,000 tonnes, 60 m high, with eight engines, a thrust of 170 m horsepower, Energiya illuminated gantries, observers and towers for miles around as it headed skywards. Energiya could put 140 tonnes into Earth orbit, perfect for putting up a huge orbital station. Indeed, one was already in design. A CIA briefing conceded that the Soviet Union now had the means to send people to Mars. They had harbored the motive for a long time.

The Mir space station was already in orbit. Two or three people were normally on Mir at any one time. At a time when the American space shuttle flew for only a week or two weeks at a time, the average Russian spaceflight was six months. Two cosmonauts were about to spend a full year on board. Fifty cosmonauts belonged to the space squad and a small group was in an advanced state of training to bring aloft, courtesy of Energiya, the large space shuttle, the Buran, which was indeed to make its maiden flight the following year.

No wonder 1987 was a year of great ambition too. *The USSR in outer space: the year 2005* drew pictures of a huge orbital station serviced by a large space shuttle. Applications satellites circled the Earth while deep-space probes set out for distant destinations. Rovers roamed the plains of Mars to bring samples to rockets that fired their cargoes back to Earth. Astronomical observatories peered to the far depths of the universe.

Buran launch

THE COLLAPSE

1987 has a surreal quality about it now. Ironically, it was the General Secretary of the Communist Party of the Soviet Union, Mikhail Gorbachev, who played an unwitting role in the program's near-collapse. Gorbachev was a keen supporter of the space program and had visited Baikonour to see Energiya just before take-off, hailing it as the achievement of Vladimir Lenin's dream of a socially progressive and technically advanced civilization. The following year, 1988, saw a party conference, where, under Gorbachev's program of openness (*glasnost*) and reform (*perestroika*), Soviet citizens had the opportunity to say what they really thought.

The old Soviet Union and the new clashed head to head the following April, 1989, when contact was lost with the second of two probes sent to Mars the previous summer. At the post-mission press conference, the old guard fought hard for a traditional explanation that the probe had been the victim of external forces ("solar storms in the Martian environment"). Most of the scientists, though, knew that the program had been mismanaged and in the spirit of *glasnost* demanded that the truth be brought out into the open. Now *glasnost* permitted many of the scientists to voice their pent-up frustration with aspects of the space program and the debate drew in a much wider audience. Public opinion, liberated under *perestroika*, turned on a secretive space program mismanaged by self-serving apparatchiks when many basic consumer goods were in short supply. During the 1989 elections, candidates pledging to cut the space budget were endorsed by the electorate. For the first time, the space program had to fight for its place on the floor of the parliament. Long-time space commentator Boris Belitsky noted: "The failure of Phobos 2 damaged space research in the eyes of the public. In the general election, several candidates proposed cuts. Failure came badly to people fed on a diet of success." In April 1990, in response, the government cut the Soviet space budget by something between R300m and R220m. This was the beginning of a long, doleful period of relentless decline.

The contraction of the space program took place in three phases. The space program had actually begun to retreat at the end of the Soviet period, in the course of 1989–90. 1989 turned out to be the peak of Soviet space spending when it stood at R6.9bn and was estimated to account for 1.5% of gross national product.

Retrenchment had most effect on the planned second flight of the Soviet space shuttle, the Buran, which had made its maiden flight automatically in November 1988. This flight was delayed until 1992, but, as it became apparent that it would likely never take place at all, workers began to leave the sheds preparing the mission at Baikonour. Then, the next planned Mars mission, Mars 94, was put back two years and other scientific projects, like the Spektr observatory, were delayed. Other planned missions to the planets and the asteroids began to disappear from the flight manifests. The annual number of launchings had already fallen from 102 a year, during the peak of the 1980s, to 75 in 1990. The number fell again, to 59, during the following year, the final one of Soviet rule.

The next contraction took place in 1992 when the Soviet Union was replaced, on 1st January, by the Russian Federation and the Commonwealth of Independent States (CIS). Within a month, the worldwide seaborne tracking fleet was recalled. The main problem was that hard currency must be paid whenever it put into foreign ports, hard currency which the country no longer had, so the ships were brought back to their home ports in the Baltic and Black Sea.

The biggest blow fell in June of the following year, 1993, when the Energiya Buran program was finally canceled. Almost twenty years of the best work of the country's leading design institutes had gone into the project, which had produced the world's most powerful rocket, Energiya, flown twice and the Buran shuttle, flown just once. The space program lost 30% of its workforce all in that one year.

1994 was the year in which the money began to come late. After a delay, most of the annual budget arrived at the year's end, leading to a surge in launchings that December. Russia's launch rate remained high, but this was deceptive, for it was due to a production line of rockets and satellites that had been paid for and built earlier. By end 1994, employment was now down to fewer than 300,000. Space spending was down to 0.23% of the national budget, compared with 0.97% spent on space research in the United States.

Newspapers forecast a further deterioration in 1995 and they were right. This time, the money that had arrived late stopped arriving in the first place. Staff in many enterprises were not paid for months. Several space enterprises were, on paper, bankrupt. Rocket launchings were held up because sub-contractors, still unpaid, would not deliver components and fuel until accounts had been settled. There was a radical reduction in the rate of military and unmanned space activities. Communications satellites which exceeded their lifetimes were not replaced. Previously, when military satellites concluded their missions, a replacement satellite was already in orbit, to make sure that coverage of targets of military interest abroad was continuous. Now, gaps appeared and the military found itself either "blind" (without photo-reconnaissance coverage) or "deaf" (without electronic coverage).

The main space magazine in Russia and one of the world's best, *Novosti Kosmonautiki*, commented in October 1996 that hardly any satellites were available to launch, and even when there were, there were no rockets to launch them. The magazine predicted that Russian space activities would probably end in a year and a half, sometime in 1997–98. This was no wild speculation, for launches that year fell to 23, less than a quarter of what they had been in the 1980s. The United States were now launching more rockets each year, something which had not happened for a generation.

Worse followed. When the much-delayed Mars 96 took off for Mars, the space program lacked the R15m needed to send the tracking ship *Cosmonaut Viktor Patsayev* to the Gulf of Guinea to follow its critical engine burn out of Earth parking orbit to Mars. The engine burn failed and Mars 96 crashed back onto the Andes Mountains. There was a half-hearted committee of investigation, half-hearted because everyone knew that this would be the last Mars probe for a very long time.

The final great contraction took place with the collapse of the ruble in the late 1990s. Now the gap between the space budget and the money that actually arrived was even wider. Only a small proportion of the money promised arrived and then at the very end of the year. Contractors in turn could not pay their suppliers and many activities slowed down or even ground to a halt. This time, there was a resigned acceptance that if government money did not arrive, it would not be coming late: it just would not be coming at all. The numbers in the space industry were now down to a quarter of the Soviet period, about 100,000. For those who remained, wages were at rock bottom: the makers of the Energomash world-beating RD-180 engine had an average wage of R3,000 a month, or $104. The military were no better off: wages in the Golitsyno control center were only R2,000 a month, or just over $60.

With a human situation like this, it was little surprise that technical quality suffered. On 5th July 1999, the normally reliable Proton rocket took off from Baikonour with a Raduga communication satellite. 390 sec later, it veered off course, debris covering a wide area and what television showed was indisputably a twisted rocket body falling into a Karaganda back garden, narrowly missing a resident 39-year-old woman and her five-year-old child. Three months later, on 27th October, another Proton exploded at 277 sec. The Kazakh government demanded €400,000 in compensation for environmental damage. At the root of it: a bad batch of engines due to poor quality control. It took months to introduce a program of quality assurance and re-qualify the errant Proton.

The decline of the Russian space program was most obvious in its physical infrastructure. In Star Town, Moscow, where the cosmonauts lived, new building plans were canceled and little or no renovation was carried out. Training facilities for Buran, including some large buildings, were left open to the elements. Things were worse in the cosmodromes. In Plesetsk there were numerous blackouts because electricity utility companies had not been paid and, to make the point, the companies would withdraw supply just as rockets were prepared for launching. Soldiers were court-martialed for selling military petrol to civilians, but their defence was that it was their only way to get long-owed backpay. In the adjacent town of Mirny, the main city bakery, renowned for its cakes, closed, to be replaced by military bread, not a delicacy!

Checking rocket engines

Baikonour cosmodrome was worse. When one of the Zenit rocket pads was destroyed in an explosion in 1990, it was simply not rebuilt. The cancelation of the Buran project led to a big exodus of personnel. There was a vacuum of power, authority and organization. There was a high level of pilferage and theft, with copper cables and sheet metal disappearing. Rubbish accumulated and was not cleared away. Air-conditioning did not work in the summer and heating failed in the winter. Food was scarce and there were weeks when basic commodities just did not arrive. As people left, schools, kindergartens and cultural facilities closed. Crime rose and people were afraid to go out at night. The rocket soldiers were mutinous and there were even accounts of riots. Energiya rockets kept indoors were guarded by soldiers in felt boots—because there was no heating in the building. Water and fuel supplies were turned off for long periods. Some parts of the city had been boarded up and were no longer occupied, launch workers being flown in from Moscow only for the periods immediately before and after the flights for which they were responsible.

The fate of Buran was one of the saddest parts of the story. The pads where it was launched were reclaimed by nature and it became no longer safe to walk there. There was no money to look after the large hangar in which it was kept and in May 2002, after heavy rainfall, it just collapsed on top of the famous space shuttle, killing eight Kazakh maintenance workers in the process. Bays 3, 4 and 5 of the huge building completely caved in. Another shuttle was bought by a joint stock company headed by the late cosmonaut Gherman Titov and relocated in Moscow's Gorky Park where 48 visitors at a time could take a two-hour, multi-media space voyage, where they could eat space food and even fend off a meteor shower. Most visitors were schoolchildren in groups: sometimes, former workers from the Buran program would come out of curiosity, but they often left in tears. Another Buran went to Australia for the Olympic Games and was never seen again.

Buran in its hangar

BACK ON MIR: THE LONG RECOVERY

The start of this chapter left Vasili Tsibliev, Alexander Lazutkin and Michael Foale in a precarious position on the Mir space station. Was there any prospect of saving the doomed station and making it operational once more?

In fact, the key decision had already been taken. Tsibliev's decision to fight to save the station—even though the manual said they should abandon ship at that stage—was the one that made the difference. Had Tsibliev, Lazutkin and Foale brought their cabin back to Earth that evening, it is hard to see another spaceship being launched to save the derelict station. Politicians were not kindly disposed to the station and President Yeltsin himself quickly accused the cosmonauts of "human error". When Tsibliev and Lazutkin eventually returned to Earth later that summer, they had a bumpy landing in more senses than one. Their final, cushioning landing rockets failed to fire, causing their Soyuz to bump badly on touchdown and they got little sympathy from mission control during their de-briefing. Their pay was suspended, bonuses canceled and they did not get the normal post-flight medals, their shabby treatment causing visible distress to their normally unflappable colleague and friend Michael Foale.

By then, although no one realized it, the Mir crisis was actually past its worst. A new crew was indeed sent up to the space station in August 1997. Pavel Vinogradov and Anatoli Soloviev repaired the station and during an internal spacewalk reconnected the cables between the Spektr module and the rest of the station. More spare parts were brought up by the space shuttle in September 1997 and the Americans replaced Michael Foale with another visiting American astronaut, NASA Administrator Dan Goldin withstanding Congressional pressure to pull American participation in the project.

Things began to settle down and, once lives were no longer in danger, Mir faded from the front pages of the world's press. Back on the ground, the causes of that summer's near-disaster were evaluated. The American conclusion, supported eventually by the Russians, was that Tsibliev had been asked to perform a series of hastily-planned maneuvers for which he had not been properly trained. Several cosmonauts were invited to perform a similar docking maneuver on the simulator in Star Town:

all crashed. In the event, Tsibliev and Lazutkin received their pay that had been stopped and their post-flight medals that had been due to them. Local fire-fighters honored them by giving them complimentary helmets. In addition, Michael Foale also received a citation for having saved the Mir station, a reward he more than deserved and the one thing that most made him feel that it had all been worthwhile.

Mir's solar panels were reconfigured and 85% of power was restored. In November, Anatoli Soloviev went on a spacewalk, holding in his hands a working model of the first Sputnik, built by Russian and French schoolchildren, transmitting on the same frequencies as the original Sputnik and delivered to Mir by a Progress freighter to mark its 40th anniversary. With a sweep of his arm, he hurled the gleaming silver ball, aerials trailing, into a separate path in orbit. In honor of the occasion, it was named Sputnik 40. But Mir was out of the news now and few media even covered the event—a sure a sign as ever that back on the world's first permanent orbiting station, it was business as usual again.

The Mir crisis had tested the Russian space program to its very limit and was probably the low point of a long, swooping line of decline. Yet, the stories of contraction, decline and crisis concealed a picture of some extraordinary ingenuity in keeping the space program going at all.

This was first evident at the very end of the Soviet period. To keep Kazakhstan on side in the space program—important, granted that Kazakhstan territory took in the Baikonour cosmodrome—a Kazakh was flown up to Mir in autumn 1991. This meant that one of the cosmonauts on board the orbiting station, Sergei Krikalev, had no seat on which to return to Earth. The solution: keep him up there. Krikalev stayed on in Mir for a further five months, coming down in spring 1992, making the longest *un*intended spaceflight ever.

Sergei Krikalev

This was only the beginning. Later missions were lengthened when rockets were not available on time. On another occasion, there was a rocket ready, but no nose cones. Soyuz launchers were borrowed from military stock to keep Mir aloft. Then, there was the sintine issue. Sintine was an expensive fuel additive used on the Soyuz U2 rocket, giving the rocket an additional kick on the way to orbit, enough to reach Mir's height. In 1996, the factory that made it in Ufa closed and the program managers turned back to the old Soyuz U rocket, which could launch slightly less into orbit. The solution to save weight: the cosmonauts had to diet and could bring only 500 g of personal effects with them.

Mir often lived on a knife edge. Although the routine flights to the station by unmanned Progress cargo craft normally went smoothly, Progress M-24 twice failed to dock. With fuel low, success was achieved on the third attempt, but if it had failed, the station would have run out of food, fuel, water and air within a matter of weeks and the crew would have to return home.

The most constructive solution was to increase the foreign, hard-currency money flowing into the space program. This at first took a sad, bizarre form. In 1993, Sotheby's auctioneers sold off 200 rare items of Russian space history, including spacesuits (Leonov's, Gagarin's and the moon suit); space cabins (e.g., Cosmos 1443); the dummy which flew on Spaceship 4; chief designer Vasili Mishin's diaries; and the slide rule of the first chief designer, Sergei Korolev (many items were bought up by entrepreneur and politician Ross Perot, who public-spiritedly put them on display at the National Air and Space Museum on the Mall in Washington DC). On Mir, cosmonauts filmed inflated Pepsi cans in orbit. Others spelled out the joys of pizzas. Veterans Talgat Musabayev and Nikolai Budarin advertized BMW cars for Spanish television. The Cosmos pavilion at the Exhibition for Economic and Scientific Achievements in northern Moscow was turned into a car salesroom. The cosmonaut training center, TsPK, offered cosmonaut training to tourists, several days costing up to $10,000, including several minutes weightless in an Ilyushin 76 trainer.

Ultimately, though, a space program could not be run on trinkets. From 1990, the space program began to sell seats on Soyuz spacecraft going up to the orbital space station. Before that, a whole parade of friendly nationalities had been brought up to orbital stations, either for free or at nominal cost. During a celebrated incident in 1975 in the course of the Apollo–Soyuz Test Project, an American journalist once pestered Intercosmos council chairman Boris Petrov to tell him how much the U.S.S.R. was putting into the project. He rambled on for half an hour. In the end he gave up, saying he didn't know. "What's the use?" he said: "I don't count the money and we still have plenty of everything we need." No more.

Now people would have to pay and pay more and more for a service hitherto given free in a spirit of socialist generosity. The first full paying visitor was the Japanese news editor Toyohiro Akiyama, whose week-long trip was paid for by the Japanese Tokyo Broadcasting System in 1990 ($12m). From 2000, an organized system of commercial space tourism was introduced, working through the American company Space Adventures, the standard price being $20m a mission. For the mean-time, the Russians sold seats to the European Space Agency (ESA) or to national space agencies in Europe or combinations of the two. These were not commercial in

Visiting mission to Mir

the sense that they were scientific missions by trained cosmonauts, but the money was real enough and equivalent to the rates charged to the tourists. From 1992 onward, there was a series of flights to Mir by cosmonauts from France and Germany, paid for by their respective national space agencies and the European Space Agency. These prices varied from $12m to around $40m, depending on the length and complexity of the mission. Originally, these payments would have gone into what might be termed the "Soviet Union general account", but as companies and corporations in the new Russia were required to be self-financing, it became payable to the Energiya Corporation which was now effectively the owner of the Mir space station.

The big breakthrough of course came with the joint Russian–American program for the Mir station. In 1993, Presidents Clinton and Yeltsin signed an agreement for the joint construction of an International Space Station which, although much of the hardware would be Russian, would be financed in large measure by the United States. In preparation for the project, there would be a joint program for the operation of the Mir station. Eight Americans would spend 600 days on Mir, ferried up there mainly by the shuttle.

The joint program with Mir set the stage for a range of international joint ventures between the leading design bureaux, production companies and scientific organizations with Western companies. Following a hastily-arranged visit by Lockheed officials to the Khrunichev factory in December 1992, the companies quickly formed a joint company to market the Proton rocket, later known as International Launch Services (ILS). They offered Proton on the world market for the launch of commercial satellites for $75m a go, quickly attracting orders. In 1995, the new revenue brought a $23m investment program to Baikonour, upgrading the airport, resurfacing the roads between the airport and the Proton pads, with improvements in the clean rooms—making the assembly and integration area one of the best in the world. In 2000 the annual production run for the Proton was increased for the first time anyone could remember.

The Lockheed–Khrunichev deal set a marker for Russia's other rockets. Lockheed's great American rival, Boeing, not to be outdone, entered a rival joint venture with the company NPO Yuzhnoye for the marketing in the West of the Zenit booster, paving the way for the Yuzhnoye/Energiya Boeing Sea Launch project (see Chapter 5). Other deals followed. Rocket engine maker Energomash sold the production rights for the RD-180 engine for $1bn to power the United States' new Atlas III and V rockets. Khrunichev sold the old Cold War rocket, Rockot, in a joint venture with the German company, Daimler-Benz. The main European deal was between the TsSKB Progress plant in Samara and the French Arianespace company in 1996, where a joint company, Starsem, was set up in July 1996 to develop and market the Soyuz rocket to launch commercial payloads. Starsem comprised four shareholders: EADS, 35%; the Russian Space Agency, 25%; TsSKB Progress, 25%; and Arianespace, 15%. Starsem provided capital to improve facilities at Samara and especially at Baikonour. The old and increasingly rundown rocket assembly hall, which dated to Sputnik, was replaced by new facilities in the Energiya assembly building. Here, Starsem invested €35m in three white rooms of world standard—a payload preparation facility, a hazardous processing facility and an integration facility—and a

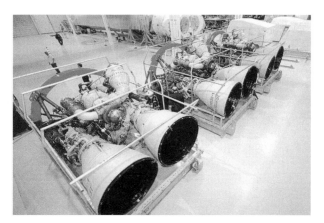

Rocket engines, an export earner

hotel for staff on site. Later, additional resources were invested in improving the two Soyuz launch pads, 1 and 31 and they were repainted. In due course, the production line for the Soyuz was increased.

ALMOST THE END: CONCLUSIONS

By 2001, 87 Russian space companies had entered joint ventures with American and European companies, leading to a visible flow of new money into the program. During the 1990s, the Russians turned their space program around from being the most self-sufficient state enterprise in the world to being the most commercial, globally connected space industry in the world. It was a difficult, painful, untidy process that has still not yet run its course. But, by the new century, the program had begun to recover. Production lines increased. New cosmonauts were recruited. In 2000, Russia once again launched more satellites into orbit than any other country and became the top space-faring nation once again. Vasili Tsibliev's role in saving the Russian space program was eventually recognized, for in 2001 he was promoted to major general on orders of the President and became commander of the Yuri Gagarin Cosmonaut Training Center.

2

Building the International Space Station

The principal focus of the Russian space program is the International Space Station (ISS), which was agreed in 1993 and started construction on orbit in 1998. Here, Chapter 2 looks not only at the origins and development of the station, but also the disposal of the flagship station fron the Soviet period, Mir. The stories of Mir, Mir 2, the American Freedom Station and the International Space Station at many times intersected each other [1].

ORIGINS OF ISS

The International Space Station had its roots in both the American and the Russian space programs. On the American side, President Reagan had, as far back as 1984, committed the United States to constructing a large space station, Freedom. By the early 1990s, Freedom had made remarkably little progress and had managed, in the course of many designs and redesigns, to consume all its budget without a single piece of hardware being built. On the Soviet side, the Russians wished to replace the space station Mir, which had been launched in 1986 with a five-year design period. Traditionally, the Russians liked to have a new station aloft before its predecessor had ceased operations and sometimes new and old stations operated simultaneously.

Originally, Mir 2 would have followed Mir: indeed, the Mir 2 service module was the backup to Mir itself, since it was normal, in case the first suffered a launch mishap, to build such modules in pairs (it still carries its 1986 hull number, #12801). When the Americans announced Freedom, the Russians reverted to their old form and scaled up the project so that it would be at least as big and ambitious. Although the planned Mir 2 module would still be used, it would be the heart of an enormous construction called OSETS or *Orbitalny Sborochno Eksploratsionny Tsentr* (Orbital Assembly and Operations Center). To the service module would be added a 90-tonne module

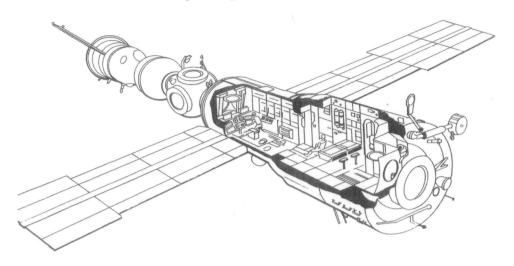

The Mir service module

launched by the Energiya rocket and huge truss structures and solar arrays. The new station would be serviced by a new spacecraft launched by Zenit rocket called Zarya and by a large Zenit-launched supply craft, Progress M2. Zarya was a 13-tonne spacecraft, to be launched on the Zenit, able to carry eight cosmonauts or four tonnes of supplies, or various combinations in between. It was shaped just like the Soyuz, but with lightweight Buran tiles as a heat shield and the downward-firing touchdown rocket moved to the sides. It was the last design of chief designer Valentin Glushko.

In the austere, post-Soviet financial climate, these plans were revised and scaled back by the Council of Chief Designers in November 1992. Mir 2 would look bigger in size than Mir because it would have an external truss structure for its solar arrays, to try and address the problems of power shortages experienced on Mir, but in reality it would be lighter. Instead of large 20-tonne modules attached to its node, the new Mir 2 would host four small seven-tonne science modules delivered by Soyuz spacecraft, a "newer, smaller, better" Mir. Because of this, it received the light-hearted, temporary nickname of "Mir 1.5".

The new "Mir 1.5" attracted the interest of the European Space Agency in spring 1993 and the project went through further evolutions to take account of European research interests. When the Buran program was canceled in June 1993, the continuation of the "Mir 1.5" version of Mir 2 was reaffirmed and a flight date was given as 1997, about the same time as Mir's termination was now scheduled. For the Europeans, Mir 1.5 could be a collaborative effort guaranteeing them a permanent place on the station, with at least one European on board at a time, ferried up by Soyuz spacecraft. The scene was set for "Mir 1.5" to evolve, over that year, as a joint European–Russian project.

At this stage, the transatlantic axis came back into play. Even as the European–Russian courtship was in progress, the United States and Russia had agreed, on 15th

July 1992, that a cosmonaut would fly on the shuttle, an American would fly on Mir, and the shuttle and Mir would link up together in 1995. The joint flights were the outcome of the much-improved political relationship between East and West that followed the Soviet withdrawal from eastern and central Europe, the reduction in size of the Russian military and agreement by both countries to no longer target their nuclear missiles at one another.

The circumstances which led the proposed but leisurely Russian–European partnership to be overtaken by a Russian–American one had much to do with the crisis around the Freedom station and the arrival of new President Clinton in the White House in January 1993. Alarmed at the costs of Freedom, he had, within a month, ordered a complete redesign by June and directed its costs to be cut from $30bn to $10bn.

In the event, NASA came back with three options: A (labeled austere), B (labeled basic) and C (which some simply called a "can"), all budgeted in a range of $12bn to $13.8bn. Congress was quite dissatisfied and a vote to kill the station there and then was lost by only one vote, 216 votes to 215. Clinton ordered NASA to come back to him in mid-September with a scaled-down version of option B, but to come in at between $10.5bn and $11bn. He also gave permission for negotiations to take place with Russia to see if cooperation were possible and if that would reduce costs.

These negotiations came to a head sooner than expected and on 5th September 1993 Vice President Al Gore and Russian Prime Minister Viktor Chernomyrdin signed an agreement. There were two parts. In the first, called phase I, the United States would pay Russia $400m for joint operations on Mir for the period to 1997. In return, Russia would provide two years of astronaut flying time on Mir. In effect, the 1992 agreement was radically extended in a purposeful way to lay the ground for a bigger project. In phase II, the American space station and the projected Russian Mir 2 space station would merge in a joint International Space Station, which would have common controls and environmental systems. The new station would be jointly assembled by the United States and Russia using the space shuttle and the Proton rocket, respectively. The European–Russian flirtation became a historical footnote, a might-have-been.

PAVING THE WAY FOR ISS: THE LAST PHASES OF MIR

Mir had never been intended to have the epic role it subsequently played in the Soviet and Russian space program. It had been conceived as a logical development of the earlier Salyut orbital stations, like which it was designed. Indeed, Mir was originally to be named Salyut 8, but political imperatives got the upper hand (Mir means "peace", or "a world where people live in harmony", the use of the word being deliberately designed to contrast favorably with President Reagan's Star Wars plans in the mid-1980s).

With a replacement to Mir not now due till the late 1990s, the Russians decided to make the most of Mir while it was available and operational, both in the joint planned program with the Americans and as a domestic space station in its own right.

Mir was occupied constantly by a resident crew of either two or three cosmonauts from September 1989 onward, generally for six months at a time. Visitors to Mir came from Japan, Britain, Austria, Kazakhstan, France and Germany, with long-duration missions by two Germans. Progress supply craft were sent up every two to three months. Extensive spacewalking was carried out, to fix experiments to the hull, take them in when they were completed and erect a series of girders, beams and maneuvering blocks on the hull. When the coup took place in the Soviet Union, mission control and the cosmonauts on board just let it blow over (whatever they thought about it, they kept to themselves).

On the experimental front, wheat was grown on board through a complete cycle. Quail chicks were hatched out. To test the idea of lighting up parts of the Earth from the sky, Progress M-15 deployed a solar sail called Znamia which cut a cone of light 5 km wide in a swathe from Toulouse through Geneva, Munich, Prague, Lodz, Brest and Gomel, where it could be seen as a bright diamond rapidly traversing the sky. Microsatellites were launched through the hatchway, like a small German satellite called GFZ with 60 geodetic reflectors. A German free-flying robot was tried out, called Inspektor.

The challenge to live with weightlessness was a routine but core concern of these missions and the cosmonauts' physical state, blood and fitness were monitored constantly. The record for the longest stay in space was extended by Dr Valeri Poliakov to 438 days, the equivalent of a trip to Mars and he was able to climb out of the capsule unaided when he returned. Elena Kondakova set a new record for a long flight by a woman, 169 days.

The American phase of Mir began with a mission colloquially called "Near-Mir" in February 1995 in which the shuttle *Discovery*, on mission STS-60, made a rendezvous with Mir. Next, in March, an American astronaut, Norman Thagard, was launched to Mir on a Russian rocket, Soyuz TM-21, the first time such an event had ever taken place. That June, the shuttle docked with Mir. Two large modules arrived: Spektr and Priroda. They had become grounded when the money ran out, but were now kitted out with American equipment and sent aloft. From 1996, the period of

Dr. Valeri Poliakov

Shuttle and Mir

American residencies began, with the first of six American astronauts on board Mir as a permanent crew member. Shannon Lucid was the first, followed by John Blaha, Jerry Linenger, Michael Foale, David Wolf and Andy Thomas.

The American missions to Mir were important for several reasons. First, they tested how the two countries could cooperate technically together and make their different designs compatible (easier than expected). Second, they gave the Americans their first experience of long-duration missions since their Skylab space station of 1973–74. Third, they provided the funding that enabled Mir to continue to function. Fourth, perhaps the least expected, was that Mir became a testing ground where two different cultures learned to work together. This actually proved to be the most difficult aspect of the missions.

The rhythm of life on orbit was a first and early noticeable difference for the Americans. With the shuttle able to make up-and-down flights for only two weeks at a time, crews worked around the clock in shifts, every moment on orbit being valuable. For cosmonauts routinely on six-month missions, such a frantic pace was neither necessary nor desirable. Cosmonauts were supposed to take time off at weekends (although this did not often happen) and try to operate a normal week. Americans were used to grabbing their meals on the move, but the Russians would

Progress approaching

insist on everyone sitting down together at the same time and taking their time, though sitting was hardly an appropriate word in weightlessness.

Some of the Americans relished the opportunity to work with the Russians on the project, Shannon Lucid and Michael Foale being good examples. Shannon Lucid's enthusiasm for the flight became apparent when she began to mail home accounts of life aboard Mir, her letters from orbit being posted on the internet for the benefit of journalists and spaceflight enthusiasts. They recounted everything from the joy of receiving a parcel of American food on board, reading who-dunnit novels in her spare time, watching the Priroda laboratory arrive ("like a gigantic silver bullet"), to sitting calmly in the airlock with her Russian colleagues as they collected their thoughts before they went spacewalking. Here, Shannon Lucid describes the arrival of a Progress supply craft:

I saw it first. There were big thunderstorms out in the Atlantic, with a brilliant display of lightning. The cities were strung out like Christmas lights all along the coast—and there was the Progress like a bright morning star skimming along the top. Suddenly, its brightness increased dramatically and Yuri said, "the engine just fired." Soon, it was close enough so that we could see the deployed solar arrays. To me, it looked like some alien insect headed toward us. All of a sudden I really did feel like I was in a cosmic outpost anxiously awaiting supplies—and really hoping that my family did remember to send me some books and candy!

Progress coming in

Shannon Lucid exercising on Mir

Soyuz and its docking system

Others, like Jerry Linenger, who found the Russians more difficult to work with, felt he had his own research program to perform and wanted to get on with it. All the Americans found the language hard and, in their own words, *bezkonechni* ("endless"). The next challenge for the Americans was learning about Mir and Soyuz systems and how the space station functioned. This they had to learn through lectures in Russian and then memorize. They could not read the manual, because there wasn't one. Although the Russians used simulators, they were a much less important aspect of training than in the United States. The Americans were genuinely taken aback at the way in which Russian training was an oral tradition, learning by doing, passed from one generation to another, without bulky manuals or instructions on a CD. European astronauts were amazed when an instructor put charts and diagrams up on the wall, pointed at them and talked about them (in Russian, of course) for an hour and then rolled them up and took them away again. No handouts: all they had was their own notes [2]. Handouts or not, three Europeans stayed long enough in Star Town to obtain a Soyuz pilot's certificate, which meant that they could take over from a Russian commander if the need arose: Claudie Hagneré (France), Thomas Reiter (Germany) and Christer Fuglesang (Sweden).

The Mir crisis in 1997 and the subsequent demands in America to "bring our boys home" created an atmosphere in which, in some parts of the press, the Russians were perceived as reckless by one side and the Americans were perceived as wimps by the other. Safety was the big dividing line between the two cultures, more at management level than at astronaut level. NASA was furious at the way in which they felt the Russians played down safety issues. The Mir collision was not the first dangerous moment, for only three months previously there had been a fire on Mir when an oxygen container had gone off and burned like a blow torch. The Russians had been in no hurry to tell the Americans about the fire, nor had they warned NASA about the dangers of the TORU experiments that brought about the collision [3].

WINDING MIR DOWN: "GRIEF IN OUR HEARTS"

With the departure from Mir of Andy Thomas, the last American, in June 1998, the way was set for the two partners to go on to build the International Space Station (ISS), for which the start was set later that year. On the American side, the view was taken that Mir should now be decommissioned: with Russia's declining financial resources, every ruble spent on Mir was a ruble that could and should be spent on getting the ISS airborne. Initially, the Russians appeared to agree. When Progress M-39 arrived at Mir on 16th May 1998, the Russian Space Agency announced that one of its objectives was to fire its engines to lower Mir's orbit from 400 km to 250 km, preparatory to final de-orbiting in 1999. A date was even given for the de-orbiting of Mir: June 1999.

At this stage, though, a campaign appeared to arise spontaneously to keep Mir orbiting. In July 1998, the Energiya Corporation made a special appeal to the Prime Minister (then Sergei Kiriyenko) to provide emergency funding to keep Mir aloft. The Russians had given commitments to fly two final fare-paying missions to Mir— one Slovak, one French—but to save money, Energiya announced that these missions would be merged.

The next move came from the Kremlin when, in an unexpected development, President Boris Yeltsin proposed that his advisor on spaceflight, Yuri Baturin, join the next Mir crew and report back to him on the condition of the station. Although then a 48-year-old lawyer on military reform, Baturin had qualified as a physicist, had worked in the Energiya Design Bureau in the 1970s and even applied for the cosmonaut squad then, though he had been rejected on grounds of poor eyesight. He was one of the main designers of the solar orientation system for Soyuz. An unusual profile for a presidential advisor.

September 1998 saw the worst financial crisis in the Russian economy since the transition to capitalism began. The ruble collapsed, there was a fresh bout of inflation and bank queues lengthened as people desperately tried to protect what was left of their savings and investments. Sergei Kiriyenko was replaced by Yevgeni Primakov, not that he made much difference. The Russian Space Agency, for its part, floated off a further 13% of the shares of the Energiya Corporation, delayed the International

Sergei Avdeev

Space Station program and slowed down the production line of Soyuz, Progress and rockets to the bare minimum.

Financial crisis or not, Mir flew on. A new crew joined Mir: Gennadiy Padalka, Sergei Avdeev and advisor Baturin duly reached Mir on the Soyuz TM-28 on 13th August 1998. Yuri Baturin relished his new assignment, made 20 experiments during his eight-day visit to Mir and duly advised his President on his return. Baturin made an unequivocal appeal for more money to keep Mir flying. He himself blended in to the cosmonaut corps effortlessly, was awarded his flight engineer's certificate in 2000 and was assigned to a new mission the following year.

Disregarding the Baturin report, the Russian Space Agency announced that the next crew to Mir would be its last. This was an assignment which no cosmonaut fought for, none wanting to be on "the crew that killed off Mir." Veteran Viktor Afanasayev was duly given command of that final assignment, set for the following year.

Routine operations continued on Mir, despite its approaching demise. Progress freighters arrived and left. The crew carried out medical, astrophysical and geophysical experiments, went on spacewalks and put another amateur handheld Sputnik working model into orbit, Sputnik 41. On 24th December, they were ordered to fire the engines of Progress, not to drop the orbit of Mir, as had been expected, but to raise it instead, albeit by a modest 14 km. This was the first indication that plans to crash Mir into the ocean in a year's time might not be so definite after all. On 1st January the cosmonauts celebrated the New Year in orbit, the ninth year in a row that orbiting cosmonauts had welcomed in the New Year. Several weeks later, new Prime Minister Yevgeni Primakov signed a decree extending Mir's life to 2002—if a foreign investor could be found. The government had apparently taken a semineutral position on Mir—not seeking its termination, but not being forthcoming in providing money either, yet prepared to indicate one possible way forward.

The crew of Soyuz TM-29 was duly launched on a white rocket on 20th February 1999, exactly thirteen years to the day since Mir itself lifted off. On board were 50-year-old veterans Viktor Afanasayev and Frenchman Jean-Pierre Hagneré, accompanied by 34-year-old Slovak research engineer and fighter pilot Ivan Bella. They were followed by what was advertized as the penultimate Progress mission to Mir on

Soyuz in orbit

2nd April. They carried out a range of experiments over the summer, made two spacewalks and followed the summer eclipse.

In June 1999 the Russian Space Agency confirmed that the current crew would be the last on Mir. The backup crew of Alexander Kaleri and Sergei Zalotin, who had been training in the hope of getting a mission to Mir in the autumn, was stood down. The station would be put in automatic mode and de-orbited in February 2000. Thus it was that at 6 p.m. on 27th August 1999 Viktor Afanasayev closed the hatch on Mir for the last time. "We have grief in our hearts," he told television viewers as he did so. The crew collected in as much as they could of scientific results and equipment. They programmed the computer to run the complex without a human presence. They closed off the fans and the ventilators and the station fell silent. Separation took place at 9:17 p.m. and Mir could be seen receding in the distance over Petropavlovsk. Soyuz TM-29 came down two orbits later 80 km east of Arkalyk. The solid rockets fired, cushioning the final descent but also setting some steppe grass on fire. The cabin turned over, making Jean-Pierre Hagneré quite sick, but he was pulled out in 15 min, given water and quickly recovered and was soon greeted by his wife and fellow cosmonaut Claudie. His colleague, Sergei Avdeev, landing with him, had been in orbit 389 days, making a personal accumulated total of 742 days in three missions. Mir continued to circle the Earth between 353 and 359 km, a level gradually falling. In December, the station was taken out of hibernation for computer and control tests. Progress M-42 fired its engines to raise its orbit from time to time.

In a last desperate effort to raise funds to keep Mir aloft, cosmonauts and engineers made a public appeal for funds. Hard-pressed Russian citizens contributed

Jean-Pierre Hagneré

what they could, but the total amount collected was only R485,000, about $15,000, enough to keep Mir going for only a few days.

It was all over. Or so it seemed until 20th January 2000 when Russian Space Agency officials confounded the critics and announced that a foreign investor had been found for Mir. Such rumors had abounded before and various names had been mentioned, from Rupert Murdoch to the government of the People's Republic of China. The unlikely savior of the Mir station was an American technology investor called Walt Anderson who ran a company called Golden Apple registered in the British Virgin Islands. He decided to invest $20m in Mir and set up a new company for the operation, MirCorp, registered in the Netherlands. Walt Anderson and his representatives talked to Energiya and they devised a plan for a 45-day mission which would refit Mir, report on its condition and prepare it for new operations while further investors were found. MirCorp spoke of how Mir could be a base for new experiments in zero gravity and even a space hotel.

Talk aside, Russia asked for a downpayment of $7m. Unlike previous promises of money, this time it actually arrived. Energiya responded quickly by assigning two Progress freighters to the MirCorp project and ordered the Soyuz TM-29 backup crew quickly back into training for a new mission. To the fury of the Americans, they sequestered a Soyuz and a Progress freighter from their slim production line of spacecraft destined for the International Space Station. The Russians just shrugged their shoulders: "We'll make good the production line later," they said.

No sooner was the cash register ringing with the MirCorp cash than Progress M-42 fired its engines again to raise the orbit, to 301–324 km. On 2nd February a new Progress, designated M1-1, took off for Mir with supplies of water, food and fuel for the station's new crew. This new version of Progress, the third of its kind, was able to carry 220 kg more cargo than its predecessor and had a total of four propellant tanks.

Progress M prepared for launch

The mark of Cassandra

Sergei Zalotin and Alexander Kaleri resumed training. The idea was that their mission would pave the way for a space film, which would involve an actor, Vladimir Steklov and an actress, either Natalia Gromushkina or Olga Kabo. Under the direction of film-maker Yuri Kara, they would make the first space drama. The film under consideration was called *The mark of Cassandra—the last journey*, by Kyrgystan-born Chingiz Aitmatov about—appropriately enough—the efforts of a Russian space commander to save a doomed orbital station. The storyline was of a renegade cosmonaut determined to stay on board an ageing space station till the end of its days—until ground control decide on the ultimate means to lure him back to Earth: send up a woman cosmonaut! Plans for their mission went through numerous evolutions. Steklov surprised everyone by passing all his training, but Kara's film backers failed to deliver the money, so he was stood down at a very late stage [4].

Despite this setback, but with sufficient money from MirCorp, Sergei Zalotin and Alexander Kaleri duly took off from Baikonour on 6th April 2000 on Soyuz TM-30 to recommission the Mir station on a mission set to last 45 days or more. On arrival on the orbiting station, which had now completed 14 years aloft, their first task was to track down a number of troublesome air leaks. Pressure in the station had fallen and they burned oxygen candles to bring it up from 628 mm. Next, they again fired the Progress engines to raise the orbit of the complex. The recommissioning of Mir took place just as the Sun was reaching the busiest point in its 11-year cycle. At solar

Alexander Kaleri

maximum, the Sun's rays had the greatest effect in agitating the molecules of the upper atmosphere, increasing air drag, which caused Mir's orbit to dip by about a kilometer each day. Hence the continued need to re-boost the orbit.

The cosmonauts managed to locate and seal the small air leak which had been bothering them—there had been a slight break in the seal on the pressure plate connecting Spektr to the node. A second supply craft, Progress M1-2, arrived on 28th April. Zalotin and Kaleri made a 4-hr 52-min spacewalk on 12th May, to inspect the hull, in the course of which they found, on the outside of the hull, a burned-out cable which had shorted. This was the 72nd spacewalk from the station. They restored all Mir's systems to full working order and carried out a full range of experiments with the orbital greenhouse (*Svet*) as well as medical and other tests. The station had never been in better working order, at least not since before the fire and collision in 1997. After a relatively uneventful 73-day mission, they undocked late on 15th June and made a smooth return to Earth 45 km southeast of Arkalyk early the following morning. Zalotin and Kaleri were removed from the cabin, helicoptered out and flown to Star Town later in the day.

No sooner were they down than MirCorp announced that it had put together a further two missions for Mir. With funding from the American Santa Monica-based, 59-year-old technology investor Dennis Tito, two long-stay expeditions would fly to Mir in 2001. Dennis Tito himself would ride up with the second crew, spend a week on board Mir and return with the first crew, leaving the new, long-stay cosmonauts on board. But it was taking some time to get this new mission together and once again Mir's clock began to run out.

A resupply craft, Progress M-43, arrived at the orbiting station on 20th October 2000. Deputy Prime Minister Illya Klebanov announced that M-43 would assist in the de-commissioning of Mir, scheduled for 27th February, to which the government had allocated €30m. But, typical of the ambiguities with which the station was viewed by the government, he added that the station might not be de-orbited if foreign investment were forthcoming. To underline the ambiguity, it was learned that Progress M-43 carried supplies that could be used for future cosmonauts. And, instead of lowering the station's orbit in advance of de-commissioning, Progress M-43 fired its engines on 25th October to raise its orbit.

The hoped-for foreign investment did not arrive and at New Year in 2001 the Energiya Corporation gave a new de-orbit date of 6th March, to be completed by a yet-to-be-launched Progress craft. Two cosmonaut teams were trained to accomplish the maneuver manually, should this be required, though how they would accomplish such a hair-raising operation was not explained.

1   Salizhan Sharipov and Pavel Vinogradov
2   Talgat Musabayev and Yuri Baturin

Later, a third crew was brought in and given the assignment. It was given the name of "expedition zero" and it replaced these four cosmonauts. Gennadiy Padalka and Nikolai Budarin had trained in 2000 as an emergency crew to fly to the Zvezda module had it failed to dock with the International Space Station and it was felt that they would have a better background for such a mission.

Expedition 0   Gennady Padalka and Nikolai Budarin

The problems of controlling an unmanned Mir became more apparent when Mir twice drifted out of alignment with the Sun, on 27th December and again on 18th January. Although control was quickly recovered, the second incident came only a couple of hours before the new Progress was due to be launched and forced a postponement of a week. Eventually, Progress M1-5 was launched on 24th January, its cargo almost entirely comprised of 2.7 tonnes of fuel for the de-orbit burn.

At this stage, Mir was losing altitude at the rate of 1,300 m a day in any case, due to the solar maximum. By March, Mir was down to a low point of 250 km, ever closer to the level at which it might make a natural reentry—but one which would threaten people down below. The danger zone was estimated to be at the 220-km mark. This was not an academic issue, for the American space station Skylab had made an uncontrolled reentry in 1979, raining debris down on the indigenous peoples of the Australian outback. Although all previous Russian space stations had made controlled reentries, Salyut 7 had not, with some components crashing into a rubbish dump in Buenos Aires. By mid-March, the rate of descent had accelerated to 1,800 m daily. Mission control followed the station very closely, as it tried to time the final burn. At TsUP mission control in Korolev, a small moving dot on a wall screen marked its progress in orbit.

The descent of Mir excited worldwide interest, for some commentators symbolizing the end of the Russian space program. It offered abundant opportunities for doom-laden predictions as to what parts of Earth might be hit if things went wrong. In Japan the Minister for Crisis Management advised people under the flight path to stay indoors. A Russian television station carried a story that Mir was full of microbiological fungi which threatened the Earth with biological catastrophe should they get through the atmosphere and mutate.

On 22nd March, Mir passed the 220-km threshold and the de-orbit maneuver was set for the following day. At mission control in Korolev, Moscow, the target was set: a 1,500-km triangle at 40°S, 160°W in the Southern Ocean, running 2,800 km east of New Zealand. On the 23rd March the fully-fueled Progress M1-5 made its first burn, 1,294 sec (21 min) at an altitude of 216 km over the Himalayas. An orbit later, the engines fired a second time for 1,445 sec (23 min) over Lake Victoria in Africa. Two orbits later, the engines were turned on over the Black Sea to burn to depletion (9 min) and by the time they finished over northwest China, the 137-tonne Mir had fallen to 159 km on a downward spiral. During this burn, a final round of television images was returned from the space station as Mir sailed high over Asia for the very last time, showing clouds and shorelines—and then the image flickered and died. Mir now entered the discernible atmosphere at 80 km at 7.47 km/sec, passing Japan on its 86,320th orbit.

Friction began to tear away at Mir and the great structure groaned and began to break up. First to snap off were the solar arrays, then the Sofora and Rapana trusses and the Strela telescopic joint. As the strains on the structure grew, the large modules broke away. Down below in the Pacific Ocean, early morning islanders gathered to see the final moments. Three plane loads of enthusiasts paid £4,500 to watch the descent from the skies. Over Fiji, streaks and streams of burning debris could be seen against the morning sky, high in the atmosphere, with palm trees in the foreground. They would glow like a large, slow meteor trail, breaking into more parts as the modules of Mir broke up and burned. There were eight distinct parts, streaks of fire followed by smoke and followed by sonic booms.

The mood in mission control in Korolev was subdued. The control room was fuller than it had been at any time since the great crisis of 1997. Space officials, "the bosses" and diplomats filled the public gallery and around them swarmed television crews from the world over. Outside, nationalist demonstrators chanted their support for the space program while communist members of parliament accused President Putin of treason. The main concern of mission control was to ensure that the de-orbit burn go smoothly. Although previous stations had been de-orbited, Mir was much larger than anything that had built before and there was always worry that radio control might be lost. Directing the proceedings was one of the first cosmonauts on Mir, Vladimir Soloviev, who tried to keep everything on an even keel. Some ground controllers could be seen wiping away the occasional tear [5].

Mission control issued a formal announcement that the Mir space station "had ceased to exist at 5:59.24 GMT on 23rd March. It has concluded its triumphant mission." A final, large image of Mir was put up on the wall screen simply listing, as on a grave, its mission dates, "20.2.1986–23.3.2001". There was no visible reaction. Controllers drifted away. Some marked the occasion with a short shot of vodka. At a press conference, Yuri Koptev made the acid comment that "although we may be very smart, we are also very poor."

Calling it a triumphant mission was not rhetoric, but the truth. Mir was the most durable single achievement of the Soviet, now Russian, space program. Mir became to the Russian space program the type of achievement that Apollo represented for the Americans. By the 21st century, over two-thirds of all time in space had been on Soviet or Russian spacecraft and most of these long hours were clocked up on Mir. There had been 27 long-stay expeditions and 69 unmanned dockings (five modules, 64 Progress and one unmanned Soyuz). Of the 11,762 days spent on board the station,

10,726 were Russian (87%), 981 American (8%) and 565 were others (5%). The period of continuous occupation was 3,640 days. Mir received 104 individuals, comprising 42 Russians (including Kazakhs), 44 Americans and 18 others. There were 28 long-stay resident missions and two visiting flights. There were nine shuttle dockings and altogether 57 visitors—from countries as diverse as Afghanistan, Austria, Britain, Bulgaria, France, Germany, Kazakhstan, Syria, Japan and Slovakia, as well as other European countries and the United States.

The Russian record in long-duration flight was, by the turn of the century, unchallenged. If one compiled a list of the longest completed spaceflights made by 2000, no American would appear on the list until the 25th name down, and that was on the Mir space station (Shannon Lucid, at 188 days, the longest traveled American astronaut). The seven longest-flying American astronauts all achieved their records on Mir. The longest flown Russian was Sergei Avdeev who spent 742 days in orbit, followed by Valeri Poliakov (678 days) and Anatoli Soloviev (638 days). One had to go a very long way down the list to meet American-launched space travelers, for the longest American flight was 84 days, set by Skylab 4 astronauts Gerry Carr, Ed Gibson and William Pogue in 1974. The longest a shuttle could fly was 17 days, still shorter than the longest solo Soyuz flight in 1970.

The volume of science carried out on Mir was enormous, problems and interruptions notwithstanding. Work done by the cosmonauts included such diverse disciplines as Earth observations, smelting, human and animal biology, medicine, radiation, antibiotics, astrophysics, atmospherics, meteorites and astronomy. Even the biology studies could be subdivided into a further range of cardiovascular studies, blood, urine, the neurovestibular system and musculoskeletal studies. Eighteen different instruments were used to measure radiation. Even some offbeat experiments proved productive: the Mariya spectrometer found a means to predict earthquakes by detecting electrostatic disturbances associated with imminent movements in the Earth's tectonic plates. One of the least acknowledged but biologically most interesting experiments was the attempt to grow plants through the entire reproductive cycle. In 1990 the Bulgarians managed to plant, grow and re-seed radish and cabbage through the whole cycle (the *Svet* experiment).

Sustaining the permanent occupation of near-Earth space required a considerable industrial and organizational commitment. The regular replacement of crews, the frequent Progress resupply missions, the year-long round-the-clock operation of flight control, the recovery of crews and capsules, the Luch communications network, all represented a commitment to quality, standards and sustained effort. The successful arrival of every manned and unmanned supply mission—100 in all—was proof if it were needed that the Russian space program could reach the highest standards of reliability. The amount of spacewalking carried out aboard Mir was enormous—it amounted to 149 spacewalks by 36 cosmonauts totaling 691 hr 18 min or almost 29 days. One cosmonaut, Anatoli Soloviev, accumulated the extraordinary personal total of 74 hr 41 min.

The demise of Mir came just forty years—short a few days—before the flight into space of Yuri Gagarin, on 12th April 1961. Space enthusiasts, dismayed but unbowed by the loss of Mir, responded defiantly by organizing the first of a series of annual

Dr Valeri Poliakov at Mir window, saying goodbye to departing astronauts

worldwide parties called "Yuri's night", to celebrate everything that had been achieved since then ("48 parties, 24 nations, 7 continents, 1 planet") everywhere from Bujumbura (Burundi) to Vancouver.

Last missions to Mir

2 Feb 2000	Progress M1-1	
6 Apr 2000	Soyuz TM-30	Sergei Zalotin
		Alexander Kaleri
26 Apr 2000	Progress M1-2	
18 Oct 2000	Progress M-43	
24 Jan 2001	Progress M1-5	

Russian space garden

BUILDING THE ISS

Under the Russian–American agreement of 1993, NASA resubmitted its new space station design in late 1993. NASA proposed four core modules: two American nodes, Mir 2 (renamed the service module) and a Russian functional cargo block called the FGB (*Funkstionalii Gruzovoi Blok*) which would be built on contract for the Americans for $215m.

The FGB was the first element of the station to fly. Although it dated to Vladimir Chelomei's Almaz space station designs of the 1960s, it was the most advanced space station of its kind when it eventually flew. The FGB was a 23-tonne module, 12.8 m long, 4 m in diameter with a volume of 55 m^3. It had two solar panels of 28 m^2, able to supply 6 kW of electricity. It had four docking ports, three at one end, one at the other. The rear end was designed to take Russia's service module, the front end America's node 1. The FGB carried two 417-kg main engines, 20 rendezvous motors of 40 kg of thrust and 16 stabilization motors of 1.3 kg of thrust. It had 16 fuel tanks holding 6.1 tonnes of propellants, 42 maneuvering engines and a tonne of micro-meteoroid shielding. Its design lifetime was a minimum of 13 years (to 2010), with at least 15 years expected. Its principal function was to stabilize the station and provide it with electricity until the station's main power system was constructed. Chelomei's ship became the core of the world's largest scientific project ever undertaken.

The FGB was the first component to launch and was to be followed by America's node 1, Unity. The FGB, renamed Zarya, was shipped from the Khrunichev factory

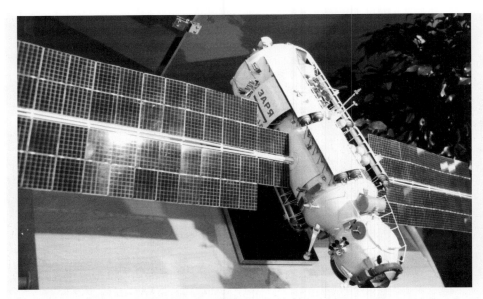

FGB Zarya

in Moscow for the 1,500-km rail journey to Baikonour in mid-winter, in January 1998. American engineers were puzzled when they learned of the extraordinarily low temperatures ($-50°C$) and high G forces (50 G) that the FGB had been built to withstand—until it was explained to them that the requirements were based on the conditions the FGB must survive on the unheated Russian railcars and on its rail and road network.

Zarya was originally to have flown earlier, but dates repeatedly slipped toward the end of 1998. The station was eventually powered up on 7th October 1998 at Baikonour, an irrevocable action committing the station to launch in the near future. NASA chief Dan Goldin stood beside Russian Space Agency leader Yuri Koptev and 160 other invited guests under a corrugated iron shelter as the Proton counted down on the morning of 20th November. Television pictures beamed the launch on the Eurovision network, showing the black-tipped Zarya against the brown steppe and a gray Kazakh sky. As usual, Russian transmissions did not show countdown clocks and because the Proton used storable fuels there were none of the telltale wisps of cold fuels burning off to indicate that a launch was imminent. Eventually, acrid yellow and orange clouds billowed out of the base of the rocket which rose with quickly accelerating speed, heading into clouds and shortly being lost to sight. At 43 km altitude, 2 min into the mission, the large first stage dropped off. At 78 km the outer fairing came off, exposing Zarya to the now airless atmosphere. The second stage was discarded at 138 km, 334 sec into the mission and the third stage then burned until 587 sec. The most crucial stage then followed: Would the solar panels deploy? After 15 min, ground control confirmed that Zarya had safely reached orbit and, to cheers, that the two 30-m solar panels had deployed.

Eventually Zarya was shipped and launched

As champagne corks popped below, Zarya was in serious trouble. The computer system on board was not responding to commands, which meant that its motors could not be fired to raise its orbit and it would decay within a few days. This problem was first picked up in the secret military tracking center in Golitsyno, west of Moscow. There, chief test engineer Nazarov wrote, in a frantic 90 minutes, a computer program to bypass the fault on Zarya and within a short period the station was responding. Zarya's initial orbit of 179–341 km was duly raised three days into its mission to 385–396 km. As for Nazarev, he received a medal from the commander-in-chief, a bonus and four years of delayed backpay!

WAITING FOR MIR 2

The next most crucial single element in the construction of the International Space Station was the Russian service module, which in reality was the Mir backup hull. Until it was ready, three shuttle servicing missions flew up to Zarya to maintain the station and extend the first building block.

The first such mission was shuttle mission STS-88, *Endeavour*, launching on 4th December, two weeks later. Besides the American crew led by Robert Cabana, Russian cosmonaut Sergei Krikalev was on board, doubly appropriate as he was scheduled to fly on the first resident Russian–American long-stay crew on the station. The first task of *Endeavour* was to attach to Zarya the first American module, called Unity. *Endeavour* used its telescopic arm to pull the 14-tonne Unity out of its payload bay and closed in on the gleaming space station on the 6th December, approaching it gingerly from above. Unity was attached and construction of the ISS was under way.

The first of three scheduled spacewalks began on the 7th December when astronauts Jerry Ross and Jim Newman left the porch of *Endeavour*, clambered onto the hull of Zarya, and began attaching 40 power cables between the two modules. Soon thereafter, ground control commanded electrical power to flow from Zarya's arrays into Unity. The astronauts installed sockets and rails to facilitate future EVAs. On a second spacewalk the next day, Ross and Newman installed two S-band relay antennas to enable the ISS to communicate with mission control in Houston via the TDRS relay system.

In true international spirit, Robert Cabana and Sergei Krikalev entered the International Space Station together on 10th December, switched on the lights and checked that all systems were in order (they were). Compared with Mir, Zarya looked fresh and uncluttered (a situation bound to change). On the 12th December, Ross and Newman made their third spacewalk, pre-positioning tools and other equipment on the outside of Zarya and Unity and making a photographic record of the conditions of both hulls. Cabana undocked *Endeavour* from the ISS on the 13th, made a lengthy fly-around and dropped away for a nighttime landing on the runway at Cape Canaveral.

The second shuttle mission up to the International Space Station took place on 27th May 1999. *Discovery* was commanded on mission STS-96 by Kent Rominger, with new cosmonaut Valeri Tokarev as the Russian visitor on board. *Discovery* reached the station on the 29th and a day later astronauts Dan Barry and Tamara Jernigan made a 7-hr 55-min spacewalk to attach Russian and American cranes to the outside of the ISS. The *Discovery* crew entered the station, transferring almost two tonnes of supplies into the station, a mixed group of items including water, clothing, sleeping bags and medical equipment. Rominger fired the engines of *Discovery* to raise the orbit of ISS by 9 km. After 80 hr linked to the station, he brought the shuttle back for a night landing. Later, it transpired that the visiting astronauts had suffered serious discomfort and nausea in the course of their visit to Zarya. Initially attributed to the air control systems in the Russian module, it turned out that the explanation lay in disrupted patterns of air flow between the station and the arriving shuttle, a spaceborne equivalent to sick building syndrome.

Originally, the third shuttle mission to the ISS was to have flown after the service module launch, but the Americans decided to bring it forward. The main reason for doing so was that four batteries on Zarya had accidentally discharged and required repair. The shuttle *Atlantis* left Cape Canaveral on mission STS-101 just before dawn local time on 19th May after a long period of countdown delays. The mission was commanded by Jim Halsell and included, as its Russian crew member, Mir veteran Yuri Usachov.

Atlantis closed in on the International Space Station two days later. Astronauts Jim Voss and Jeff Williams soon suited up for a 6-hr spacewalk, attaching a new

Russian Strela crane and reattaching some external equipment that had come loose. They placed a range of external equipment on the outside of the station. The astronauts opened the hatches into Unity and Zarya on 22nd May, becoming the first visitors to the complex in a year. They rigged up a new system of air supplies, recharged the FGB batteries and then used the *Atlantis* engines in 27 pulses of 130 sec each to raise the station's orbit by 30 km.

DELAY, DELAY AND DELAY

Next to fly was the Russian service module. Although the American-financed FGB was completed in 1997 on schedule (though launched late), the Russian service module languished due to lack of Russian funding and ground to a halt. The hull, Mir's spare, had been completed as far back as February 1985 and the module fitted out by October 1986. However, its internal systems had to be completely rebuilt and modernized so as to operate with the FGB and the other components of the ISS.

Repeated American visits to the Energiya plant in Moscow produced renewed commitments to resume work, but nothing ever seemed to happen. Frustration grew and NASA began to consider contingency plans should Russia drop out of the project. Not until May 1997, following the personal intervention of President Yeltsin, did funding resume on the Russian side with a presidential commitment of $260m. Within days, engineers were back at work on the project. However, this did not lead to the expected improvement, possibly because President Yeltsin's new funding turned out to be a complex set of bank loans rather than on-the-spot cash. Dates again slipped and slipped. The situation worsened with inflation and the slide of the ruble on foreign exchanges. In 1998, as launch dates came and went, the only progress was that the service module acquired a name, Zvezda, or Star. Reluctantly, in late 1998, the Americans provided more money, between $60m and $100m in order to try and resume some momentum. This was traded against NASA's expectation that Russian crews on the ISS would contribute time to American experiments.

The new funding was linked to a plan whereby the assembly of the station would be stretched over another year, to 2005, and the shuttle would make up for some Russian resupply missions. The issue was a tricky one for NASA: many in Congress and the press were unhappy about the links with the Russians in any case and were quick to exploit these bailouts. Even those who favored the payments worried about the way in which money seemed to disappear when it arrived in Russia. There were allegations that top officials were pocketing the money and building fancy private homes for themselves in Star Town. NASA did not wish to buy more Russian parts of the ISS outright, like the FGB and indeed Congress might well have prevented this. In the end, the formula was that NASA rented future Russian facilities on board the station, even though they had not yet flown. NASA spent 1998 investigating a number of scenarios whereby it could substitute its own module for the service module and proceed without Russia. An interim control module, ICM, was even designed. Reluctantly, NASA concluded that a substitute module would cause further delays, do much less and ultimately cost more. It was cheaper, albeit frustrating and even exasperating, to continue to subsidize the Russians.

By 1999, these additional injections of cash had their effect. By then too, the financial crisis had eased a little and the service module was at last shipped to Baikonour. It might have flown that autumn had not the Proton rocket which was due to launch it suffered a double launch failure in July and October. The Russians insisted (as did Kazakhstan, which was likely to suffer environmentally the most from further disasters) that there be a careful, step-by-step requalification program. Zvezda was slipped to May 2000, then July. The Proton returned to flight on 12th February 2000 and Zvezda was finally powered up. A surge of Proton launches in the early summer cleared the way for the crucial service module launch.

For the United States, these delays were testing, although the Americans glossed over problems on their own side which would certainly have led to some (though much less serious) delays to the project. Many Americans attributed these delays to the costs and energy involved in keeping Mir aloft. The sooner the old station was de-orbited and sunk in the Pacific Ocean, the sooner the ISS project would be under way, they argued. When the Russians recommissioned Mir in spring 2000, the Americans responded by announcing that they would build their interim control module if the Russians failed to orbit Zvezda by the autumn. There was even an implied threat in some quarters to kick the Russians out of the ISS project. Relationships between the two sides were poor in spring 2000. The General Accounting Office presented a report listing the defects of the Russian FGB, such as loud ventilators, poor air supplies and insufficient protection against depressurization—even though the module had been built to American specifications and so certified before launch.

ZVEZDA, 12TH JULY 2000

Zvezda counted down in the scorching summer of the Baikonour morning heat of 12th July 2000. Just to remind people of the cash shortage, advertising space on the side was sold to Pizza Hut, who paid €1m to Energiya for the privilege of relaying its logo to the world's watching television companies. Against a light-blue sky, brown clouds billowed out beside the lightning towers on the Proton pads as Zvezda sped skyward. Although unmanned, it was probably the most important single launch of the program, for the prospects of replacing a failed Zvezda were financially not worth thinking about. At 80 km, halfway through the second-stage burn, the payload fairing was jettisoned and Zvezda started to transmit its own data.

Just less than ten minutes after take-off, Zvezda was in orbit 184 km high and 3 min later its 30-m solar panels sprung open, beginning a 13-day rendezvous profile. NASA's Dan Goldin put his previous comments to one side and spoke of all the difficulties the Russians had overcome to launch the station "and they came through." Ground control in Korolev then steered Zvezda through a series of maneuvers, bringing it into an ever-higher orbit toward the FGB. At 80 km distance from one another, the approach and docking maneuvers were begun, with Zvezda being the passive ship. The FGB was commanded to yaw and made a radar sweep for

Zvezda and Zarya together, at last

Preparing a Progress M

its sister ship. The whole process was flawless and by early morning on the 25th July the two craft were 200 m apart, 366 km over Kazakhstan. The FGB moved gently in at 0.2 m/sec and they clunked together.

Over the next week, ground controllers in Korolev and Houston began reconfiguring the software of the FGB, Unity and Zvezda so as to achieve a unified control system. On 6th August the first unmanned supply craft for the ISS, Progress M1-3, left Baikonour for the station to bring up equipment to prepare the station for permanent occupation, arriving with food and water, oxygen generators and personal computers two days later. Progress M1-3 closed in in darkness, using a floodlight to illuminate Zvezda for the last 150 m. Once it docked, the station comprised three modules and one supply craft, weighing 60 tonnes. On 10th August, Progress fired its engines to raise the height of the station by 2.8 km.

Just as Zarya was the essential block that controlled the International Space Station, Zvezda was the module that would make it possible for astronauts and cosmonauts to live and work there. Like its predecessor, Mir, it had no scientific equipment, but it hosted bedrooms, a bathroom, running track and exercise area, cooking galley, refrigerator and even an entertainment rack with videos, CDs and books. The men and women on board could observe the Earth through 13 portholes. It carried a set of computers called the Data Management System (DMS-R), developed by the European Space Agency. The new Zvezda telemetry system, called Regul, had flown only once before, on Buran and was now making its operational debut. Many of the bulky consoles of Mir had been replaced by new slimline, portable computer systems. In a small but welcome home improvement, the Russians brought a space-age samovar on board—a dispenser able to serve hot tea at any time, day or night.

The first expedition to the International Space Station comprised Mir veterans Yuri Gidzenko and Sergei Krikalev and shuttle pilot William Shepherd. Professionally, they crossed a range of disciplines from MiG pilot (Gidzenko) to underwater commando (Shepherd) to Energiya engineer and computer genius (Krikalev). Their first scheduled task was to get the toilet and air systems in working order, unpack the scientific and other equipment on the space station, unload new supplies arriving on Progress M1 spacecraft and welcome two shuttle missions—one to deliver the second American module, Destiny, and the other to bring up the first solar array.

At last, after all the preparations, the first permanent crew was launched to the International Space Station on 31st October 2000 [6]. It was a sign of the new Russia – or arguably, the very old—that a black-frocked orthodox priest was on hand to sprinkle holy water on the crew members before they dressed in their spacesuits. Visually, the take-off was not much of a spectacle, because the white, orange and gray rocket disappeared into low cloud almost as soon as it was launched, but television provided a live relay from the cabin as it sped skyward. The three-person crew arrived at the station on 2nd November, settled in and made a live hook-up to the Kremlin where they were congratulated by President Putin. The head of NASA, Dan Goldin, proclaimed that from this day on there would always be people in space.

There was a hiccup two weeks later when the unmanned supply craft Progress M1-4 arrived on 18th November. For the first time, it was to dock with a docking port on the FGB, but the automatic docking system, Kurs, failed to lock on. Progress began to oscillate and rock, its movement clearly visible on the television relays to mission control in both Moscow and Houston. Yuri Gidzenko now took over with the TORU system, problems with which had led to the Mir collision three years earlier. Gidzenko tried to guide the Progress in, but he found it extremely difficult, due to fogging on the screen and poor lighting conditions compounded by reflected sunlight. Despite this, he managed to bring Progress M1-4 a mere 9 m from the docking port, at which point the station entered darkness and no one could see anything at all. At this stage, mission control in Moscow suggested he back Progress off to 35 m for the half-orbit of darkness before trying again. After a patient wait, he tried again 45 min later and at 4 m distance the fogging problem suddenly cleared, giving him a clear run to complete a smooth docking. It was evidently a one-off problem, for the next Progress arrived smoothly at the end of February.

William Shepherd, Yuri Gidzenko and Sergei Krikalev returned to Earth at the end of the first expedition to the ISS on 20th March. They came back to Earth, not as they had left, in the Soyuz, but on the American shuttle, because it had been decided that, the start of the first mission excepted, all crew changes would be undertaken by the American space shuttle. The expedition 2 crew was brought up by the shuttle that brought them down: Yuri Usachov, Susan Helms and James Voss. The pattern was to alternate two Russians–one American with two Americans–one Russian. Meantime, Soyuz missions to the ISS would continue, but they would fulfill a different role.

SOYUZ AS LIFEBOAT

From the beginning, both countries defined a need to be able to evacuate their crews speedily should an emergency take place, a well-identified need in the light of the Mir fire and collision in 1997. The United States had planned to develop a lifeboat for the

Yuri Usachov

space station, a spaceplane called the X-38, but when the space station budget overran, this was canceled. In the meantime, the Soyuz had to do.

It was decided that from April 2001 a Soyuz would be permanently attached to the space station, ready to bring the crew home in an emergency, such as a fire, depressurization or loss of control. A problem here was that the Soyuz had a guaranteed on-orbit lifetime of 180 days, after which it was considered that its systems would decline and fuels corrode. This meant that a fresh Soyuz must be brought up to the space station every six months and the old one brought down. This decision set the scene for the six-monthly launches of Soyuz to the space station and from then have generally taken place every April and October, or March and September.

Soyuz was almost used as a getaway during the crash of 1997, although the cosmonauts stopped the air leak just in time. It came to light, thanks to the work of a Swedish researcher in 1999, that the Soyuz always had a procedure for making a quick return to Earth [7]. As far back as 1979, the Kettering Grammar School group of amateur radio sleuths in England had listened in to transmissions from Soyuz 33 and heard the crew referring repeatedly to the *ugol pasadki* or landing angle (literally in Russian, "landing corner"). Later, cosmonaut Vladimir Shatalov confirmed that every night, before a crew went to sleep, it set the landing angle so as to be able to make an emergency descent at any time. The *ugol pasadki* was, to be precise, the angle between retrofire, the Earth's center and the landing spot. Analysis of Soyuz landing angles found that the Russians had designated a series of emergency landing zones worldwide. Within the former USSR itself, there were several emergency return spots: the Sea of Okhotsk, Poltava and near Kiev in the Ukraine. However, the most intriguing landing areas are those outside the Russian Federation. These are worth listing, since the inhabitants of these areas have probably been blissfully unaware that, for years, they have been a Soyuz emergency recovery zone! In North America, these are Manitoba, Saskatchewan, North Dakota, Forth Worth (Texas), Oklahoma, all presumably selected because of their flatness and similarity to the steppes of Kazakhstan. In Europe there are two selected zones: the plain of Flanders (Belgium)

Soyuz landing

and Picardy (France). So, in the event of the crew of the International Space Station having to bail out, these are the points where they may come down.

In the course of building the ISS, the Russians introduced a new version of the Soyuz, called the Soyuz TMA. This was the fourth main version of the Soyuz spacecraft to be introduced since 1966 after Soyuz itself, Soyuz T (Transport) and Soyuz TM (Transport Modified). The TMA (Transport Modified Anthropometric) featured improved computers, better controls, a new and smaller control panel, new life support systems and more physical room. During the missions to Mir, it was found that about half America's astronauts were too tall to fit into the old Soyuz—a question of being the right stuff, but not the right size. The anthropomorphic Soyuz TMA provided the extra few centimeters to accommodate them (190 cm, against 182 cm) and extra weight, 95 kg against the previous maximum of 85 kg. Control

The new Soyuz TMA

panel units not necessary for ascent and descent were moved below. A new air-conditioning unit was installed. The shock absorbers were modified. The soft-landing rockets were adjusted to make landings smoother. The altimeter was designed to fire the rockets more accurately. The TMA was 80 kg heavier and required a new launcher, the Soyuz FG rocket.

SPACE STATION ROUTINE

Starting 2001, the International Space Station settled into its routine of:

- three-person expedition crews, changed by the American space shuttle every 180 days or so;
- construction visits by the American space shuttle to bring up fresh modules and equipment;
- external construction and experiments carried out by astronauts and cosmonauts on spacewalks or EVAs (extra vehicular activity);
- Soyuz lifeboat changed every 180 days, normally spring and autumn;
- unmanned resupply by the Progress cargo craft, between two and four a year.

Typically, astronauts slept 8.5 hr, worked 6.5 hr and exercised 3 hr a day. Saturday and Sunday were supposed to be days off (except for exercise, which is never optional), though this rarely happens.

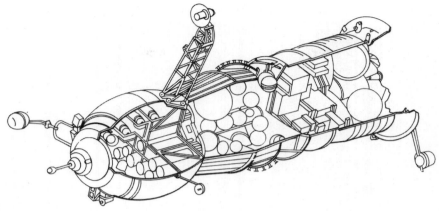

Progress diagram

The resupply of the space station was quite a challenge. Sustaining three astronauts on the space station required a continuous stream of supplies of food, water and fresh clothes, as well as experiments and equipment to repair items that broke down. Over 2000–02, a series of American shuttle flights brought supplies and new equipment up to the ISS. One of the most efficient ways of doing so was to use three small Italian logistics modules (named *Leonardo*, *Raffaello* and *Donatello*) which were brought up in the shuttle cargo bay, docked to the space station, unloaded, reloaded and brought back again.

As well as the "up" role played by the shuttle and Progresses, the station generated a considerable amount of "down" rubbish. In the early days of space stations, this had been pushed out an airlock, but in the course of time this was realized to be unecological as well as dangerous, for the debris could strike the station when their orbits intersected in the future and cause damage. Accordingly, the shuttle had an important role in bringing rubbish back to Cape Canaveral, while departing Progress spaceships were filled with non-returnable rubbish before they were de-orbited to destruction away from the shipping lanes in the Southern Ocean.

Progress spacecraft were not always de-orbited immediately on undocking. For example, in September–October 2002, Progress M-46 spent 19 days on an independent Earth resources mission. Using its engines to raise its orbit to 401 km, it was used to film forest fires in Russia and floods in eastern Europe. Similarly, Progress M1-10 made a month-long environmental monitoring mission from 412 km the following year. Progress M1-11 in May 2004 spent two weeks testing orientation systems for future microgravity missions, as did Progress M-51 early the following year [8]. The most unusual one was Progress M1-7 which in March 2001 deployed its own sub-satellite, the 22-kg Kolibri, or humming bird, an Australian–Russian project to monitor solar plasma storms. Kolibri orbited for 45 days, relaying information back to ground stations in Omsk and Sydney before burning up on 3rd May.

There was always the danger that the Progress resupply missions might seem mundane and become taken for granted. Not so in December 2004. As a result of

Progress launch sequence

Ready

Ignition

Liftoff

Climbing

Afterwards

some form of household under-budgeting, or the hearty appetites of previous crews, the crew of Salizhan Sharipov and Leroy Chiao found itself short of food coming up to the end of 2004. The two men had to eat much less volume to conserve existing supplies, but thankfully there was a plentiful supply of chocolate bars to keep them going. Both lost weight, about 2 kg to 3 kg each. Progress M-51 brought up much-needed relief on the 25th in the form of 200 kg of 70 food containers with 67 days of full rations and 45 days of 80% rations. There was only five days food left at this stage and had the Progress not arrived, then the station would have had to be evacuated by the end of the year—illustrating the thin margin on which it was operating at the time.

Unloading Progress was one of the favorite activities for the crews on the station. Before it arrived, the crew would go to the windows to spot the little spacecraft closing in against the blue and white of the Earth and the black of the night. They would watch its engines fire as it came in and hear the dull thud as it arrived. Sometimes, the ISS crew used the TORU system to guide the Progress in manually. Despite the difficulties of the first use of TORU in 1997, which led to the Mir collision, TORU turned out to be an easy system to use. American astronaut Ed Lu recalled that the main problem was that it took a second or two for the command on the joystick to actually turn into an event, but once you got used to the delayed reaction, it was not difficult. Most times the Progress came in automatically and it was not needed.

Unloading generally took about three or four days, because everything had to be bolted firmly down. The crew loved to get the fresh fruit, simple things like apples and tomatoes, but such a change to canned food. They received letters and pictures from their families, who normally also sent up some compact disks and videos. Then the time would come to start filling up the Progress with rubbish, like old clothes and food cans, but being careful to log everything and make sure it had a stable center of gravity.

Progress missions often brought unusual "up" cargoes. Progress M-52, for example, carried fifty snails to test their reaction to weightlessness, to come down on the next returning Soyuz. Progress spacecraft were also frequently used to raise the orbit of the space station: for example, an 831-sec burn in April 2003 lifted the orbit from 380 km to 400 km and then later to 420 km. In November 2005, Progress M-54 made a continuous 33 min burn to raise the height of the orbit. Later, the American science officer Peggy Whitson recalled how she barely noticed the start of burn, but she soon began to drift through the station in a very straight line.

The building of the ISS has been described well in other places [9]. From the Russian point of view, the highlight of the first manned phase of operation was the arrival of the Pirs ("pier") docking unit. This, the third Russian module to the ISS, arrived on 17th September 2001. Unlike the large Zarya and Zvezda modules, which were launched by Proton, Pirs was a small module, 3.6 tonnes in weight, similar to Progress, launched on a Soyuz U. Essentially, it was a docking port and an airlock module, like a ship's pier, volume 13 m^3, with two hatches for spacewalks and handholds to enable the astronauts to begin their work. Pirs included one Russian Orlan M spacesuit (two had been sent up to the ISS at the very beginning), just as the American Quest airlock had a set of American spacesuits, while the crew had the

Science Officer Peggy Whitson in Russian Sokol spacesuit

The Pirs module

Pirs arriving

option of using either. Pirs made a smooth flight up to the station, docked after two days and twelve latches secured the module.

Until the arrival of Pirs and Quest, ISS spacewalks had generally been conducted from the space shuttle when it was docked to the space station. With Pirs and Quest, the astronauts and cosmonauts had a base from which to conduct their regular program of spacewalks and construction activities. The airlocks enabled the crew to pre-breathe in advance of spacewalks, for the station was pressurized at 1 atmosphere, the same as Earth's atmosphere at sea level. Pre-breathing and subsequent re-adaptation were important, otherwise the crew risked the danger of "the bends", or decompression sickness, suffered by divers who make too swift ascents to the surface.

On 8th October 2001, Vladimir Dezhurov and Mikhail Tyurin made the first spacewalk from Pirs, by chance also the 100th Soviet/Russian spacewalk to be made [10]. Cosmonauts (or astronauts) exiting from Pirs used the Orlan M spacesuits, the latest evolution of a type of spacesuit that dated to the Salyut space stations in the late 1970s. The Orlan M weighs 112 kg and has an endurance of 9 hr.

Russian spacesuits are quite different from American ones, for one climbs into them through a door in the back. They are effectively a spaceship in their own right, with a one-size-fits-all approach and intended to be used by people of any size. Only the gloves are individualized. Orlan M came to be extensively used on the station, both by Russian and American crew members. By April 2006, there had been 18 spacewalks using Orlan suits (two each time, making a total of 36 Orlan EVAs). American suits, by contrast, are customized. Astronauts found them less tiring to use:

Salizhan Sharipov spacewalking

this was partly because they worked at lower pressure, 0.3 atmospheres, but this carried a greater risk of "the bends" than the Orlan's 0.4 atmospheres. The pre-breathe period for Orlan was 30 min, but 4 hr for the American suit. The Orlan was quick to get into and a cosmonaut could put the Orlan on and close the door in the rear without assistance, all in less than 5 min.

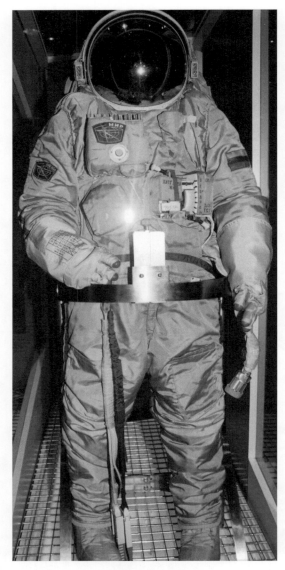

Orlan spacesuit

The ISS made Sergei Krikalev the most traveled cosmonaut in history. When he flew on Soyuz TMA-6 in 2005, his sixth mission, he overtook the previous record for duration space travel, 742 days, held by his colleague Sergei Avdeev and built up a personal total of 803 days. That was set to increase even further, for in 2006 Sergei Krikalev was named commander of Soyuz TMA-14. Another kind of record was set by cosmonaut Yuri Malenchenko. Before his mission, he had become engaged to be

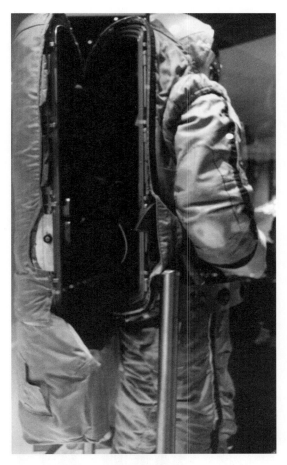

Orlan spacesuit rear view

married, with the wedding set some time ahead, for August 2003. When the time came for him to take his vows, he was in orbit and the couple saw no reason for distance to stand in their way. They were duly married through a television link to American mission control in Houston. Russian ground control was not so amused at this distraction and built into the contracts of future cosmonauts a clause prohibiting future space weddings.

Occasionally, the space station served as an orbital launching base. In March 2005, Salizan Sharipov launched—by hand—a 4.5-kg satellite called TeKh, built by the Scientific Research Institute for Space Engineering. In February 2006, Valeri Tokarev and William McArthur jettisoned, in the course of a spacewalk, an old and expired Orlan M spacesuit. As part of an experiment with amateur astronomers, it was fitted with a radio transmitter and they were able to follow its ever-weaker broadcasts until the battery ran out.

Orlan spacesuit side view

Some parts of being on the space station were not routine. The flight path of the space station brought it over a clear and sunny New York morning on 11th September 2001 and the crew was able to observe the horror of the fire arising from the attack on the World Trade Center twin towers, following the plume of smoke and ash over Manhattan. During that day, mission controllers in Houston had to evacuate to a bunker (there was a fear of more attacks) and ISS control was passed over to Moscow for two days.

Then, there was 1st February 2003. Following a 16-day independent mission which had no connexion to the space station, the American space shuttle *Columbia* was returning to Earth. It didn't make it. Because of damage to its wing during take-off, hot gases entered the left side of the wing and *Columbia* burned to destruction

Salizhan Sharipov with TeKh nanosatellite

during reentry over east Texas. Rick Husband, Kalapana Chawla, Laurel Clark, Ilan Ramon, David Brown, Michael Anderson and William McCool were lost, all persons known to the American crew on board. They watched the uplink of the memorial service, ran the ship's bell seven times in honor of their friends, observed a 20-min silence and then went back to work [11].

SPACE STATION DEPENDS ON RUSSIA

The shuttle was grounded. From now on, the space station was dependent on Russia to stay flying. Some time later, NASA Administrator Sean O'Keefe was to refer to this as a dark period, "but Russia stepped up to the plate and kept us going."

The loss of *Columbia* had no immediate effect on the pattern of work on the International Space Station, but it did in the medium term. The grounding of the shuttle fleet, which most people estimated would last for two years—in fact, it was slightly longer—meant that the supply system to the station was much diminished and fresh construction came to a halt for the time being. The two space agencies tackled the problem in the following ways.

First, the crew on the station was reduced, from three to two, diminishing the need for supply. From April 2003, the permanent, resident crew would be down to two—one American, one Russian. A downside was that a smaller crew would limit the amount of time available for experiments.

Second, the station became dependent on Russia for Progress resupplies for several years. A large European cargo craft, the Automated Transfer Vehicle (the

Jettisoned spacesuit

first was called *Jules Verne*) was in preparation, but its launch was some way off. The new version of Progress, the M1 series, introduced in 2000, was soon suspended and the old version, the Progress M, given a new lease of life. Progress M had the advantage of being able to hold more drinking water than the M1 version. The shuttle had always brought up the station's drinking water, but with the shuttle no longer available, this became a premium deliverable item. Although it had been planned to replace the Progress M when it reached the Progress M-44 mark, the Progress M series now continued indefinitely. The Progress missions were generally routine. One which caused temporary worry was Progress M-58 in October 2006. One of the rendezvous antennae of the Kurs system called the 4AO VKO failed to close in properly during the final stages of docking, causing an obstruction and the risk that a hard docking could not be achieved. The TORU system was activated to complete the docking. In the end, the station crew was able to retract the 4AO VKO and re-deploy the hooks on the ISS to obtain both a docking and a fully airtight seal.

The third change was an altered role for the Soyuz. Hitherto, Soyuz had been launched twice a year to bring up a fresh spacecraft to act as a lifeboat, the old one being brought down after 180 days when its on-orbit lifetime expired, the crew flying on week-long, up-and-down missions. Now, the permanent crew of the space station would have to go up on a new Soyuz, the returning one bringing the previous one down. Thus, the next Soyuz, TMA-2, brought up the next two-person expedition crew, Yuri Malenchenko and Ed Lu. They left the third seat empty, because the returning Soyuz TMA-1 had to bring down the previous three-person expedition crew of Donald Pettit, Kenneth Bowersox and Nikolai Budarin.

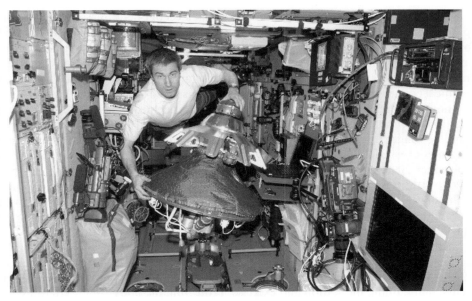

Sergei Krikalev with Progress docking system

Cosmonauts set out for ISS

Although the Russians provided a substantial level of operations to and from the space station, a significant level of cost was paid for by the United States. NASA paid for the Soyuz seats bringing its astronauts up to the orbital station. That agreement ran out in 2005 and then fell foul of the international political situation. Under the Iran non-proliferation legislation passed by Congress, restrictions were placed on technical cooperation with Russia unless Russia was given a clean bill of health that it was not transferring technology to Iran. The Russians denied that such transfers were taking place, but this was contested. As a result, it looked as if NASA astronauts could no longer fly on the Soyuz. If shuttle missions did not resume, this meant that no Americans could fly to and from the station any more. Congress relented in November 2005 and authorized NASA to negotiate seats on the Soyuz from April 2006 to 2012. The shuttle did not return fully to flight until July 2006. The restoration of the shuttle schedule also made possible the resumption of three-person crews, with the addition of Thomas Reiter of Germany on a long-duration mission for the European Space Agency.

Daily life on the space station is now well known, thanks to the accounts of the astronauts and cosmonauts on board. Several e-mailed their experiences back to Earth. Astronauts and cosmonauts divided their time between carrying out experiments, maintaining the station and exercise. Protecting the body from weightlessness was a dominant activity. Cosmonauts were expected to exercise twice a day for up to 2 hr, using the bicycle, running track and expanders. It was boring too; so, as in Earthly gyms, the crew members sometimes watched videos to while away the time. Exercise was especially important before going on a spacewalk or in the two weeks

Gennadiy Padalka prepares for spacewalk

before returning to Earth. Spacewalking was always strenuous—cosmonauts expected to lose about 2 kg to 3 kg in weight as a result—so it was important to be fit. The Institute for Medical and Biological Problems (IMBP) in Moscow kept a close watch on whether counter-measures were being taken and was critical of crews that failed to exercise properly [12]. Some things were easier though, such as sleeping where, because of weightlessness, there were no pressure points on the body and you didn't need to turn to get a better position.

Most of the initial spacewalks from the International Space Station were carried out by American astronauts spacewalking from the space shuttle while docked to the station and in connexion with construction activities. Not until September 2000 was the first Russian ISS spacewalk and that was also from the shuttle before the station received its first resident crew. The first Russian spacewalk from the station itself was one undertaken by Yuri Usachov with James Voss in June 2001. Not until the third ISS expedition in autumn 2001 was a significant level of Russian spacewalking undertaken from the ISS: this followed the arrival of the Russian airlock module, Pirs, that September. Altogether, there were 23 spacewalks by Russian cosmonauts, most from the Pirs module on the International Space Station.

Russian spacewalks 2000–6

Date	Cosmonaut	Companion	Exit point	Duration
12 May 2000	Alexander Kaleri	Sergei Zalotin	Mir	4 hr 52 min
10 Sep 2000	Yuri Malenchenko	Ed Lu	Atlantis	6 hr 14 min*
8 Jun 2001	Yuri Usachov	James Voss	ISS Zvezda	19 min
8 Oct 2001	Vladimir Dezhurov	Mikhail Tyurin	ISS Pirs	4 hr 58 min
15 Oct 2001	Vladimir Dezhurov	Mikhail Tyurin	ISS Pirs	5 hr 52 min
12 Nov 2001	Vladimir Dezhurov	Frank Culbertson	ISS Pirs	5 hr 4 min
3 Dec 2001	Vladimir Dezhurov	Mikhail Tyurin	ISS Pirs	2 hr 46 min
14 Jan 2002	Yuri Onufrienko	Carl Walz	ISS Pirs	6 hr 3 min
25 Jan 2002	Yuri Onufrienko	Dan Bursch	ISS Pirs	5 hr 59in
16 Aug 2002	Valeri Korzun	Peggy Whitson	ISS Pirs	4 hr 25 min
26 Aug 2002	Valeri Korzun	Mikhail Treschev	ISS Pirs	5 hr 21 min
26 Feb 2004	Alexander Kaleri	Michael Foale	ISS Pirs	3 hr 55 min
24 Jun 2004	Gennadiy Padalka	Michael Fincke	ISS Pirs	14 min
30 Jun 2004	Gennadiy Padalka	Michael Fincke	ISS Pirs	5 hr 40 min
3 Aug 2004	Gennadiy Padalka	Michael Fincke	ISS Pirs	4 hr 30 min
3 Sep 2004	Gennadiy Padalka	Michael Fincke	ISS Pirs	5 hr 21 min
26 Jan 2005	Salizhan Sharipov	Leroy Chiao	ISS Pirs	5 hr 28 min
28 Mar 2005	Salizhan Sharipov	Leroy Chiao	ISS Pirs	4 hr 30 min
18 Aug 2005	Sergei Krikalev	John Phillips	ISS Pirs	4 hr 58 min
7 Nov 2005	Valeri Tokarev	William McArthur	ISS Quest	5 hr 22 min*
3 Feb 2006	Valeri Tokarev	William McArthur	ISS Pirs	5 hr 43 min
1 Jun 2006	Pavel Vingradov	Jeffrey Williams	ISS Pirs	6 hr 31 min
23 Nov 2006	Mikhail Tyurin	Michael Lopez Alegria	ISS Pirs	5 hr 38 min

*American suits
Sources: Shayler, 2004 [10]; Corneille, 2005 [10]; Joachim Becker: www.spacefacts.de

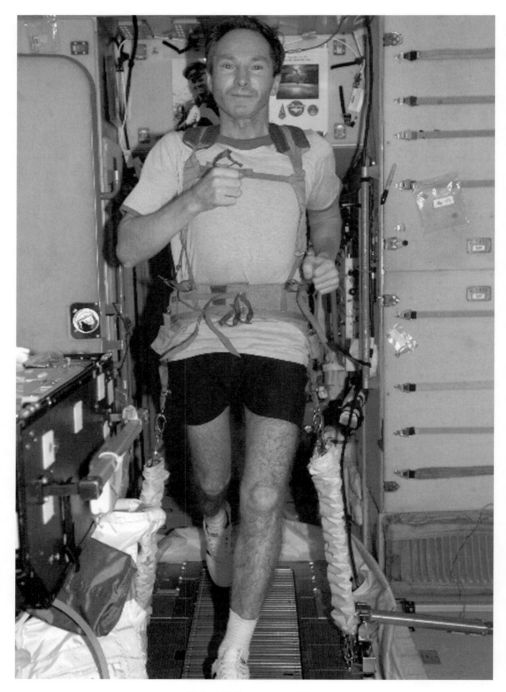

Valeri Tokarev exercising

The Russians went to considerable length to monitor the medical conditions of their cosmonauts in orbit and to combat the worst effects of weightlessness. No fewer than four Russian–American boards supervised medical aspects of the flights. There is a daily medical check and report, with in-depth checks using medical equipment at intervals (e.g., blood, electrocardiogram tests). Spacewalks, because of their strenuous nature, require an especially high level of pre-walk and post-walk monitoring and rest is often ordered after them to recover from fatigue. Radiation, of course, was checked constantly. Levels of radiation ranged from 1,865 millirads for the first crew to 3,850 on the twelfth [13].

One unexpected medical issue they had to watch for was microbiological infection. Ever since mould had appeared on one of the early orbit stations (Salyut 4), the biologists had become aware of the capacity of orbital stations to host bacteria and fungi. They formed both a health threat to the cosmonauts and had the capacity to degrade equipment. An inventory on Mir logged 148 species of bacteria and 151 species of fungi! [14]. For the ISS, there was a higher level of environmental monitoring, using air and chromatograph sampling, which found 70 microbial species, but levels were within limits. One of the surprises was that the main source of the bacteria was not bugs that had been living in the modules for years, but arriving cargo ships and visiting crews which brought with them a flood of fresh bugs and contaminants. A different problem on the ISS was noise, which was persistently 4 to 16 decibels above the prescribed limit, forcing the cosmonauts to wear ear plugs, ear mufflers and acoustic ear flaps.

The two sides, the Americans and Russians, always got on well together while on orbit. At one stage, the two space agencies were arguing about who should control what part of the mission, Moscow or Houston, but the crew on board simply got on with it and told the two ground controls to let them know when they had sorted it out. Crew members kept in touch with each other even after missions ended and they had returned to their respective homes in Star Town and Houston.

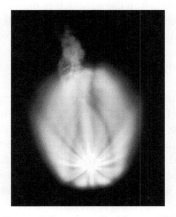

Soyuz FG rocket climbing to ISS

Still, differences of approach were evident. Americans were very conscious of doing things by the manual, whereas on the Russian side cosmonauts learned about systems from lectures by the designers who made them. The Americans were taken aback that the Russians never wrote anything down. Yuri Usachov, on the second expedition crew, once faced the problem of having to change the large batteries on the station: "doing it by the book would have taken two to three days longer," he said "and it was much easier just to get on with it."

A substantial amount of science was carried out on the Russian part of the space station [15]. Many of these were the descendants of experiments begun on the Mir station. Some science experiments were installed on the station from the very beginning, but others were brought up by Progress spacecraft and the American shuttle. Russia defined 331 experiments to be carried out on the ISS. These were as follows:

Life science research	70
Technology	53
Biotechnologies	47
Geophysical	36
Earth's natural resources	32
Materials science	19
Power supply systems	14
Astronomy of solar system	10
Radiation	8
Atmospheric science	6
Combined	36

The total weight of the experimental packages was set at 26 tonnes. This program of research was well under way by 2006, although the amount carried out was still much less than that achieved on the Mir space station (Table 2.1). A basic problem on the ISS was the station itself required a maintenance crew of two astronauts and they had little time left over for scientific work. Improved utilization could only be achieved with a doubling of the space station crew, which is planned over the 2008–10 period.

A total of 149 experiments were carried out on the Russian segment over 2001–06, the principal foreign countries being Italy (27), the Netherlands (24), Spain (20) and Belgium (17). One of the big problems was the difficulty in getting experimental results back to Earth, for Soyuz had only 50 kg of cargo space for its return missions [16].

Shortage of funding continued to dog Russian participation in the space station. Although the severity of the financial crisis receded in the new century, it continued to recur. In late 2003 a Progress launch was put back because of lack of funds. That year the government had voted an additional R2.8bn to support ISS operations, but it still did not arrive by year's end [17]. Such problems continued intermittently.

The ISS continued to be a test-bench for long-duration missions for the planets. A mundane example of this was the Russian effort to develop water recycling: clearly, Progress resupplies would not be possible en route to Mars. A cosmonaut requires 2.5 liters of water each day. On the old space station, Salyut 4, Alexei Gubarev and

Table 2.1. Selection of Russian science experiments, International Space Station

Area	Name and type
Human life sciences and medical	Reaction to weightlessness Heart performance—electrocardiograms Muscle performance-dynamometer Analysis of blood, urine and air Fluids in the human body (*Sprut*) Dental health in orbit (*Paradont*) Cardiac activity (*Kardio*) Effectiveness of medicines (*Farma*)
Plants, cells and biology	Greenhouses: *Lada, Rasteniye, Svet* Fruitfly genetic studies (*Poligen*) Bioreactors (*Rekomb K, Konjugatsiya*)
Space environment	Effect of meteorites on the station (*Meteoroid*) Magnetic levels outside station (*Iskazhenie*) Microbes inside the station Dynamic loads on ISS (*Identifikatsiya*) Acoustic environment (*Infrazvuk*)
Cosmic rays and radiation	Effects of rays on outside and inside (*Platan*) Radiation levels (*Prognoz*) Radiation and biology (*Bradoz*)
Space environment	Testing for hypergolic residues, contamination and propellant residue (*Kronka*)
Earth observations	*Uragan*—observation of the Earth for hurricanes, volcanoes, disasters, signs of human activities Ocean blooms (*Diatomeya*), with research ship *Mstislav Keldysh*
Geophysics	Lightning (*Molniya*) Seismic activity Glows in the upper atmosphere (*Reaksiya*)
Materials-processing	Plasma crystals: *Plazmenniy krystall*
Proteins and crystals	*Glycoproteid, Mimitek*

Georgi Grechko had tested the SRV-K system which took condensation out of the air and turned it into water. On Mir, the SBK-U system was developed to turn urine into water and on the ISS the Russian section carried the 115-kg SRV-K2M system (condensation) and 75-kg SPK-VM system (urine). By 2006, the Russian systems had between Salyut, Mir and the ISS produced 6.3 tonnes of water in space, water which did not have to be hauled up by Progress freighters.

The main problem on the ISS has actually been in this area. To generate fresh oxygen, the ISS was equipped with an oxygen generator in Zvezda called Elektron. Its

Yuri Malenchenko with plasma crystal experiment

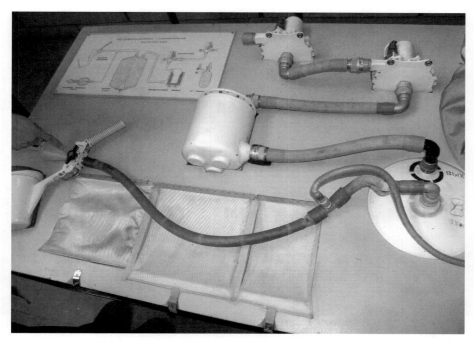

Urine removal system on ISS

role was to take waste water, separate oxygen from hydrogen, dump the hydrogen and make fresh oxygen for the crew. Simply put, it kept breaking down and replacements did not seem to work either. Although not life-threatening, the problems with Elektron used up a completely disproportionate and frustrating amount of crew time.

COMPLETING THE SPACE STATION

The International Space Station is already the largest object ever assembled in the sky. Down below, people on Earth living between 51°N and 51°S (most people) are able to spot the station crossing the night sky at its set height of 444 km as a bright and steady object, its solar panels glinting in the Sun. From time to time they are able to see the regular procession of shuttles, Soyuz and other ferries flying up to the station, bringing up supplies and men and women and their equipment.

The total/airtight volume of the International Space Station will reach 1,200 m^3. The mass of the station, when complete, will be over 415 tonnes in weight, 74 m wide and 108 m long, as big as two football fields end to end. The array of solar panels along its truss will generate up to 110 kW of electrical power. The overall cost of the project, from 1993 through to 2012, is likely to be in the order of €40bn, making it

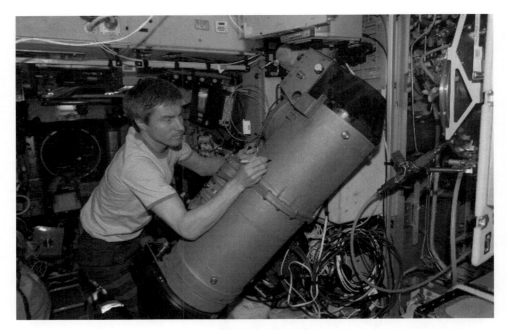

Sergei Krikalev with Elektron system

the largest international scientific project ever undertaken. Although led by the United States and the Russian Federation, there is a substantial investment by Europe, Canada and Japan. Europe and Japan are supplying modules and unmanned cargo ferries, while Canada is supplying mobile equipment and maneuvering arms.

Russia's third large module was originally to be the Universal Docking Module. The term "docking module" was something of an understatement, for it was much more than that. The UDM was to be a 20-tonne module, 14 m long and 4 m wide, with 80 m^3 of habitable space, an airlock, three docking ports for Soyuz and Progress modules and the segment connecting the Russian part of the station to the rest, with its own power supply, life support systems and some research equipment. Although manufacturing began in 1999, little progress was made and two years later Russia accepted that it did not have the money for the project. Russia was also expected to provide a science and power platform for the ISS and this too went through many evolutions, becoming smaller each time [18]. Because of its awkward shape, this was slated for launch on the American space shuttle.

Instead, Russia decided to see what could be done with the backup version of the FGB, called FGB 2. This had been constructed, at Russian expense, alongside the American-financed FGB and would have flown if the launch of Zarya had failed. FGB 2 had the advantage that the hull had been built and fitting it out was 70% complete. Plans to finish and launch the FGB 2 went through numerous evolutions

Soyuz coming in

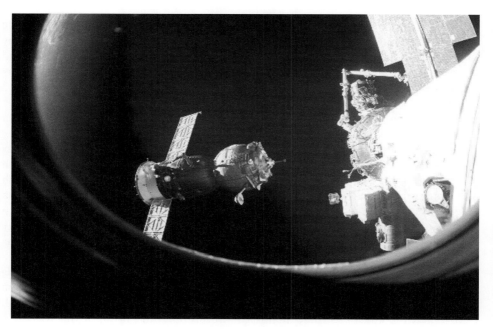

Soyuz—final stage of docking

over the following eight years. At one stage, it was planned to furnish FGB 2 as a commercial module called *Enterprise*, but this foundered in a mixture of political problems.

Eventually, in 2004 the FGB 2 was renamed the Multi-Purpose Laboratory Module (MPLM) and it had a broadly similar function to the intended UDM. The MPLM will include a 11 m long European robotic arm built by Dutch Space in Leiden, the Netherlands, one of the most sophisticated robotic arms ever attached to a space station, able to lift eight-tonne modules while at the same time having a dexterity of 5 mm. The MPLM will be attached to the FGB 1/node 1 junction and for clearance reasons must be in place before the third station node arrives. The MPLM weighs 20.3 tonnes at launch (close to 24 tonnes on completion), generates 3 kW of electrical power from its solar panels, has an internal volume of 71 m^3 and a docking port for Soyuz or Progress. It is also expected to improve comfort on the station, for it will carry a sleeping room, sauna and additional toilet. The MPLM will have several workstations for cosmonauts on the outside and on the inside, room for three tonnes of scientific experiments, 8 m^3 of storage and a workstation whence to operate the Dutch arm.

Although the FGB 2 had been built by Khrunichev, the Energiya company was in 2006 awarded the contract to finish the module and deliver it to orbit in 2009, followed, on the last shuttle flight, by the Scientific Energy Module (*Nauk Energiya Modul*—NEM) in 2010 and then a Research Module (*Issledovatl Modul*—IM) in 2011.

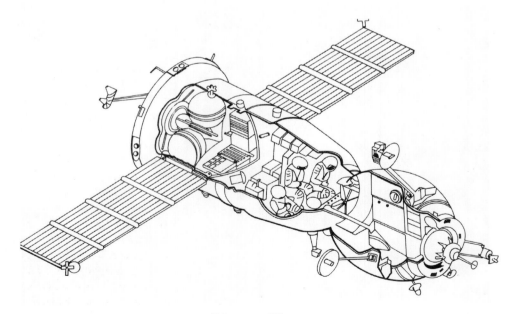

Diagram of Soyuz

Both would be installed at the junction between Zarya and Zvezda. The NEM was a module with eight solar arrays able to generate 5 kW of electricity, with storage batteries and gyrodynes to orientate the space station. The IM will in effect be FGB 3, with 13 m^3 of space for scientific experiments.

In 2005, Russia announced that it would increase the production rate of Soyuz and Progress from the then rate of two Soyuz and four Progress a year to four Soyuz and eight Progress a year from 2009. This would enable a six-crew operation of the space station. Instead of a Soyuz launch every April and October, there would be two spring launches about a month apart and two such autumn launches. With the retirement of the space shuttle in September 2010, there will be a four-year gap before the next American manned spacecraft flies, the new Orion, which looks like the old Apollo. Soyuz will therefore be the sole means of accessing the station from 2010 to 2014.

The original manifests committed Russia to 122 launches between 1997 and 2012, the main elements of which were Progress resupply missions (73) and manned Soyuz spacecraft (32). As such, this was a much higher level of commitment than the United States (79 launches) and which was considered internationally as the leader of the project. By 2000, in the light of Russia's financial problems, revised manifests had reduced that commitment to ten Proton assembly missions, 25 Soyuz and 30 Progress M, 65 spacecraft in all. By end 2006, Russia had actually launched to the ISS two Protons, one docking module, 23 Progress (up to Progress M-58) and 13 Soyuz (including Soyuz TMA-9) (39 spacecraft).

The future of the space station after its completion is uncertain. The Russians found the Americans to be difficult partners at times and the Russian Space Agency occasionally sketched plans to develop its own national space station circling at 65° to the equator, an inclination much better for observing the Russian landmass. It was never clear whether this was a serious plan or whether it was done to remind the Americans that Russia could still go it alone. All this changed when President Bush announced, in 2004, the Vision for Space Exploration, with his plan for returning Americans to the moon and then flying onward to Mars. Although Bush gave a commitment that the United States fulfill its obligation to complete the International Space Station, he had little to say about a continuing American involvement there. Although the new American manned spacecraft, the Crew Exploration Vehicle, would visit the space station, it was clear that American interests would be refocused elsewhere. It was difficult to avoid the conclusion that once shuttle missions ended in 2010, the International Space Station would become, ever more, a Russian-led project, developed in cooperation with Europe, Canada and Japan. The more this looked likely, the less we heard about Russian intentions to launch a Russian-only space station at high latitude.

Russian launches to the ISS: modules
20 Nov 1998   Zarya
12 Jul   2000   Zvezda
15 Sep 2002   Pirs

Russian manned launches to ISS, 2000–6

31 Oct 2000	Soyuz TM-31	William Shepherd
		Yuri Gidzenko
		Sergei Krikalev
30 Apr 2001	Soyuz TM-32	Talgat Musabayev
		Yuri Baturin
		Dennis Tito (Space Adventures)
21 Oct 2001	Soyuz TM-33	Viktor Afanasayev
		Sergei Kozeyev
		Claudie Hagneré (ESA/France)
25 Apr 2002	Soyuz TM-34	Yuri Gidzenko
		Roberto Vittori (ESA/Italy)
		Mark Shuttleworth (Space Adventures)
28 Oct 2002	Soyuz TMA-1	Sergei Zalotin
		Yuri Lonchakov
		Frank de Winne (ESA/Belgium)
26 Apr 2003	Soyuz TMA-2	Yuri Malenchenko
		Edward Lu
18 Oct 2003	Soyuz TMA-3	Alexander Kaleri
		Michael Foale
		Pedro Duque (ESA/Spain)
19 Apr 2004	Soyuz TMA-4	Gennady Padalka
		Michael Fincke
		André Kuipers (ESA/Netherlands)
14 Oct 2004	Soyuz TMA-5	Salizhan Sharipov
		Leroy Chiao
		Yuri Shargin
15 Apr 2005	Soyuz TMA-6	Sergei Krikalev
		John Phillips
		Roberto Vittori (ESA/Italy)
1 Oct 2005	Soyuz TMA-7	Valeri Tokarev
		William McArthur
		Gregory Olsen (Space Adventures)
9 Apr 2006	Soyuz TMA-8	Pavel Vinogradov
		Jeffrey Williams
		Marcos Pontes (Brazil)
18 Sep 2006	Soyuz TMA-9	Mikhail Tyurin
		Michael Lopez-Alegria
		Anousheh Ansari (Space Adventures)

All on Soyuz rocket from Baikonour: Soyuz U for TM series, Soyuz FG for TMA series

Progress launches to ISS, 2000–6

6 Aug 2000	Progress M1-3
16 Nov 2000	Progress M1-4
26 Feb 2001	Progress M-44
20 May 2001	Progress M1-6
20 Aug 2001	Progress M-45
26 Nov 2001	Progress M1-7
22 Mar 2002	Progress M1-8

26 Jun	2002	Progress M-46
25 Sep	2002	Progress M1-9
2 Feb	2003	Progress M-47
8 Jun	2003	Progress M1-10
29 Aug	2003	Progress M-48
29 Jan	2004	Progress M1-11
25 May	2004	Progress M-49
11 Aug	2004	Progress M-50
24 Dec	2004	Progress M-51
28 Feb	2005	Progress M-52
18 Jun	2005	Progress M-53
8 Sep	2005	Progress M-54
21 Oct	2005	Progress M-55
24 Apr	2006	Progress M-56
28 Jun	2006	Progress M-57
23 Oct	2006	Progress M-58

All on Soyuz U from Baikonour, except M1-6, 7, 8 and 9 which took Soyuz FG

The following is a listing of the expedition crews to the International Space Station. The first expedition crew was launched by the Soyuz (TM-31) and subsequent crews by the shuttle until the *Columbia* disaster in February 2003. Following this, the crew was reduced in size to two, but launched again by Soyuz (TMA-2).

The expedition crews to the International Space Station

No.	Crew	Launch	Landing	Duration (days)
1	William Shepherd Yuri Gidzenko Sergei Krikalev	31 Oct 2000	21 Mar 2001	141
2	Yuri Usachov Susan Helms James Voss	8 Mar 2001	22 Aug 2001	167
3	Frank Culbertson Vladimir Dezhurov Mikhail Tyurin	10 Aug 2001	17 Dec 2001	129
4	Yuri Onufrienko Carl Walz Daniel Bursch	5 Dec 2001	19 Jun 2002	196
5	Valeri Korzun Peggy Whitson Sergei Treschev	5 Jun 2002	7 Dec 2002	185
6	Kenneth Bowersox Nikolai Budarin Donald Pettit	24 Nov 2002	4 May 2003	161
7	Yuri Malenchenko Edward Lu	26 Apr 2003	28 Oct 2003	185
8	Michael Foale Alexander Kaleri	18 Oct 2003	30 Apr 2004	195

The expedition crews to the International Space Station (*cont.*)

No.	Crew	Launch	Landing	Duration (days)
9	Gennady Padalka Michael Finke	19 Apr 2004	24 Oct 2004	188
10	Leroy Chiao Salizhan Sharipov	14 Oct 2004	25 Apr 2005	193
11	Sergei Krikalev John Phillips	15 Apr 2005	11 Oct 2005	179
12	William McArthur Valeri Tokarev	1 Oct 2005	9 Apr 2006	190
13	Pavel Vinogradov Jeffrey Williams	30 Mar 2006	29 Sep 2006	184
14	Mikhail Tyurin Michael Lopez-Alegria Anousheh Ansari	18 Sep 2006		

BUILDING THE INTERNATIONAL SPACE STATION: CONCLUSIONS

The International Space Station was an opportunity for Russia to continue to maintain its commitment to keeping cosmonauts in space. It is very doubtful if Russia could have, without the ISS, afforded on its own to launch the Mir 2 space station, crew it and resupply it. American cash provided a means whereby Mir 2 could be kitted out, the FGB built and the production line for the Soyuz and Progress kept rolling.

On the other hand, there was a debit side to the space station. Financial dependance on the United States for funding at key points created an unhealthy relationship in which Russia came to see itself, and was seen internationally, as the junior partner. Indeed, the International Space Station came to be projected in some of the Western media as an American project in which the Russians were either minor players or were being given a minimal role out of charity. American financial dominance meant that money talked whenever there were, as indeed there were, arguments about such diverse matters as crew composition and the measurement system to be used during the program (metric or imperial). This did less than justice to the enormous legacy of orbital construction which the Russians brought to the project, the very real and tough level of Russian financial investment (€10bn over its twelve-year base lifetime) and the considerable number of launches and resupply missions to which the Russians were committed, 39 being carried out over 1998–2006. The International Space Station was heavily dependent on Russian expertise, its large modules (Zarya, Zvezda) and the fact that it provided an average of half the permanent crew, either two or three cosmonauts at any given time. Although the Russians had no option but to participate in the ISS in 1993, it was often forgotten that the Americans had been equally obliged to do so too, for political support for an independent go-it-alone American station had ebbed by the same time.

Although the United States were the main bankers for the International Space Station, the Russians provided substantial resources of their own. Of the initial $21bn assembly costs (not the total cost of the project, which is expected to be $40bn) of the ISS for 1993–2005, $13bn was borne by the United States, Russia $2.51bn, Europe $2.46bn, Japan $2.36bn and Canada $68m. The Russian contribution was small compared with the United States', but it was still the second largest and for an economy in Russia's state, this was nevertheless a huge and stressful investment. When the shuttle was grounded, it was Russia that kept the station going, a fact always generously acknowledged by NASA.

Despite its difficult start, Russia had managed to build its part of the International Space Station and sustain its operations, launching thirteen manned spacecraft to the station, as well as 23 Progress resupply missions. What about its unmanned space programs?

REFERENCES

[1] For a detailed account of the origins and development of the ISS, see Hendrickx, Bart: From Mir 2 to the ISS Russian segment, from Rex Hall (ed.): *The International Space Station—from imagination to reality*. London, British Interplanetary Society, 2002.

[2] William Triplett: Astronaut, cosmonaut ... Euronaut? *Air & Space*, August/September 2003.

[3] Oberg, Jim: *Star-crossed orbits*. McGraw-Hill, 2002.

[4] Vis, Bert: Mir crews—the final year, from Rex Hall (ed.): *Mir—the final year*. London, British Interplanetary Society, 2001.

[5] Zak, Anatoli: Fallen star—the day the station died. *Air & Space*, June/July 2001; Clark, Phillip S: Mir—the final nine months in orbit, from Rex Hall (ed): *Mir—the final year*. London, British interplanetary Society, 2001; Harland, David: *The story of space station Mir*. Springer/Praxis, 2005.

[6] Dickey, Beth: The captain, the pro and the fighter pilot. *Air & Space*, February/March 2000.

[7] Grahn, Sven: Ugol pasadki, *www.svengrahn.ppe.se* 2005.

[8] Clark, Phillip S.: The International Space Station—orbital considerations and related topics January 2002–April 2005, from Rex Hall (ed.): *The International Space Station—from imagination to reality*. London, British Interplanetary Society, 2005.

[9] Bond, Peter: *The continuing story of the International Space Station*. Springer/Praxis, 2002; Kidger, Neville: The International Space Station—a site under construction, from Brian Harvey (ed.): *2007 Space Exploration annual*. Springer/Praxis, 2006; Vis, Bert: ISS crewing, from Rex Hall (ed.): *The International Space Station—from imagination to reality*. London, British Interplanetary Society, 2005; Kidger, Neville: The expedition crews—life in orbit from Rex Hall (ed.): *The International Space Station—from imagination to reality*. London, British Interplanetary Society, 2005.

[10] Shayler, David J.: *Walking in space*. Springer/Praxis, 2004; Abramov, Isaak P. and Skoog, A. Ingemar: *Russian spacesuits*. Springer/Praxis, 2003; Corneille, Philip: ISS EVA log, 1998–2005, from Rex Hall (ed.): *The International Space Station—from imagination to reality*. London, British Interplanetary Society, 2005.

[11] Thomas D. Jones: The first 1,000 days. *Air & Space*, June–July 2004.

[12] Kozlovskaya, Inessa and Grigoriev, Anatoli: *Russia system of countermeasures on board the International Space Station—the first results*. Paper presented to the 54th conference of the International Astronautical Federation, Bremen, Germany, 29 September–3 October 2003.

[13] Grigoriev, Anatoli; Bogomolov, Valery; Goncharov, Igor; Alferova, Irina; Katuntsev, Vladimir; and Osipov, Yuri: *Main results of medical support to the crews of the International Space Station*. Paper presented to International Astronautical Federation, Valencia, Spain, 2006.

[14] Grigoriev, A.I. and Pestov, I.D.: *Russian approaches to the medical care systems for extended spaceflights—experiences and challenges*. Paper presented to International Astronautical Federation, Valencia, Spain, 2006.

[15] Salmon, Andy: Research in orbit, from Rex Hall (ed.): *The International Space Station— from imagination to reality*. London, British Interplanetary Society, 2002; Russian science on the International Space Station—paper presented to the British Interplanetary Society, 7th June 2003.

[16] Sorokin, Igor & Markov, Alexander: *Status and problems of space station utilization*. Paper presented to International Astronautical Federation, Valencia, Spain, 2006.

[17] Kislyakov, Andrei: Space station on the verge of financial collapse. *Spaceflight*, vol. 45, December 2003.

[18] Hendrickx, Bart: The ISS Russian segment—recent developments and future prospects, from Rex Hall (ed.): *The International Space Station—from imagination to reality*. London, British Interplanetary Society, 2005.

3

Scientific and applications programs

Manned spaceflight has always been the most dramatic, visible part of the space program. Unmanned programs, though less glamorous, have often brought greater benefits to ordinary citizens, be that through communications, weather forecasting or navigation. During the heyday of the Soviet space program, the Russians ran a broad range of unmanned programs, ranging from communications to weather satellites, missions with monkeys to materials-processing, Earth observations to space science. With the financial contraction of the 1990s, these programs were inevitably reduced in scale. Here we look at how Russia managed to adapt these programs to a very different financial environment. We start with the original Soviet applications program: communications satellites.

COMSATS: THE SOVIET INHERITANCE

Russia was a poor second in the communications satellite (comsat) field after the United States. The Americans first experimented with relaying messages through satellites as early as 1958. Soviet comsat technology tended to lag about five years or more behind the United States and this gap never really closed. First, we look at the communications satellites inherited from the Soviet period, like Molniya, Raduga, Gorizont and Ekran, before going on to examine the new-generation, Russian-period comsats.

Molniya (the Russian word for "lightning") was the USSR's first communication satellite. Molniya was a cone shape. At the base, six vane-like panels spread out like a windmill about to turn. Each weighed about 816 kg and was 4 m high, 1.4 m in diameter. Like all the other Soviet comsats, Molniya was built by the NPO PM Design Bureau in Krasnoyarsk.

Molniya became the basis of the Soviet comsat system. It used a system of orbits not hitherto used by any other type of satellite system in the world—but one easily explicable in terms of the geography of Siberia. Most communications satellites use 24-hr orbits, 36,000 km out called geosynchronous orbits: they circle the Earth once a day over the equator and, because their orbit matches the Earth's rotation, they appear to hover over the same point constantly. This is ideal for stable communications and three or so such satellites can make a global system. But, in designing a comsat system for Russia, the 24-hr orbit presented difficulties. A stable 24-hr orbit can only be located on the equator and a transmitter from equatorial orbit is stressed in trying to reach Russia's northerly latitudes, especially the far north. At the other extreme, a conventional orbiting satellite in a low orbit covering Siberia would present major tracking problems, because it would be overhead for periods of only ten or fifteen minutes at a time.

Hence the compromise 12-hr orbit stretching up to the 24-hr altitude of 36,000 km, but with a perigee of 600 km. Apogee was normally over Siberia, so that the satellite climbed slowly across that part of the sky over a period of eight hours, before whizzing around its perigee at a much faster velocity. One disadvantage of this orbit is that it takes the satellite through the Van Allen radiation belt and, as a result, each satellite requires some heavy shielding.

The idea of using such an orbit for communications had first been presented by Bill Hilton in the British Interplanetary Society in 1959–60. He had sketched the idea of such a high-eccentricity orbit in 1960. He may have had a bigger audience than he realized, for Hilton was immediately accosted by an enthusiastic Aeroflot official wishing to discuss the paper with him. The paper was at once translated into Russian and the Molniya system went into design the following year.

The first attempted Molniya launch was in 1964, but it failed and success was not achieved till April 1965. A series of 12-m receiver dishes called Orbita 1 was built throughout the USSR to pick up Molniya signals and by the 1980s over a hundred such dishes had been built. Molniyas originally operated in a constellation of eight at a time, thus ensuring continuous coverage. They were developed in three main series—1, 2 and 3—but with subsets.

Ninety-one Molniya 1s were built and flown by 2000, with two launched since then (1-92 in April 2003 and 1-93 in February 2004). A subset of Molniya 1s received the "T" designation to mark the use of a different system of transponders, starting in April 1983 with Molniya 1-57, although the existence of the T subset was not made known for eight years. Sometimes, the "1" and "T" designations were run alongside one another, with the danger of confusion but without a full list of the T series being formally available. Thus, the Molniya launched in April 2003 was known both as Molniya 1-92 and also 1T-28.

An improved version, Molniya 2, appeared in 1971, but was quickly replaced by Molniya 3, which first appeared in 1974. Molniya 3 was able to use much smaller ground stations and by the early 1970s its terminals were as small as 7 m (the Orbita 2 system). Like Molniya 1, the 3 series operated in constellations of eight. At a general level, Molniya 1 concentrated on domestic, governmental and military communications links and Molniya 3 on international and civilian links.

The Molniya 3 series concluded on 19th June 2003 with Molniya 3-53, the manufacturer announcing that this would be the last one of the "3" series. A modernized version, the 3K series, was introduced in 2001 with Molniya 3K, the first being on 25th October 2001, designed to extend the working life from three to five years. To continue the confusion of running two designators alongside one another, it was also called 3-52. Thus, in the period 2000–06, five Molniyas were launched (see list at end of this section).

Although the Molniya system formed the core of the Soviet communications satellite system, the USSR also developed a system of 24-hr communications satellites. Once again, the Americans were first off in the mark and their relay satellite, Early Bird 1, introduced global television from 24-hr orbit in 1965, the Russians ten years later. Three types emerged: standard TV and communications satellites, Gorizont (the Russian word for "horizon"), which were civil; Raduga satellites (the Russian word for "rainbow"), which were government and military; and direct broadcast satellites (called Ekran, the Russian for "screen"), which were civilian.

Gorizont (industry code 11F662) looked like an automaton out of science fiction, with arrays of shutters, engines, tanks and instruments atop a cylinder base and a huge dish, pointing earthwards. A typical Gorizont weighed 2.5 tonnes and carried 11 antennas and eight transponders. Gorizont ran as a civilian television and telephone transmission satellite from 1978 and concluded with Gorizont 33 in 2000.

Its military and government equivalent was Raduga, which had both a first generation (industry code 11F638, called *Gran*, although the term was little used) and a second generation (17F15, called *Globus*). Although no picture of Raduga has been published, it is believed to look the same as Gorizont. They are believed to carry satellite telephone lines for both the military and for government departments and agencies. By 2000, there had been 33 Radugas (the first generation had concluded) and four in the improved Raduga 1 series. Older Radugas continued to operate. For example, Raduga 1-4, put up in 1999, did not retire until the end of 2005.

Three Radugas were launched in the new century, Radugas 1-5, 1-6 and 1-7. Whereas the Molniya series operated a confusing set of numbers, announcements about these Radugas simply stated that a launch had taken place in the Globus series. These Radugas may mark the phasing out end of the series, as a new generation of domestic communications satellites becomes available, like Ekspress, from which the government could lease services. Raduga 1-6 retired early in 2006.

Ekran (industry code 11F647), introduced in 1976, carried television only and was one of the first national television systems designed to broadcast direct to the homes of isolated communities. The target area was a footprint covering Novosibirsk, Irkutsk and northwest Mongolia. All that people needed on the ground was a small dish rooftop aerial. This precision delivery to scattered communities was made possible by the unusual design of the Ekran: a cylinder pointing earthwards. On top were two solar panels stretching from a long beam. Underneath was a flat electrical board and, spread out on it, 200 pencil-shaped antennas pinging downwards. Ekran weighed two tonnes and had a lifetime of over two years. The government provided about 5,000 initial receivers for the region, where they were designed to serve an initial target audience of 20 million people. Ekran continued the tradition of putting more and more power on the satellite and less and less receiving power on the ground, hence ever-smaller dishes. Ekran had a broadcasting power of 200 watts,

Ekran

compared with a mere 8 to 10 watts on the Raduga. The service as a whole eventually covered 40% of the USSR, or $9\,\mathrm{m\,km}^2$.

Like Raduga, there were two generations of Ekran. The first Ekran series was brought to a conclusion with Ekran 17, but an upgraded version, the M series (industry code 11F647M), was introduced in 1987 as Ekran M-1. Confusingly, Ekran Ms were known both under Ekran and Ekran M designators and an Ekran could have both an Ekran number and an Ekran M number. Only one was launched in the new century, bringing the series to a conclusion, Ekran M-4. Ekran M-4 was destined for the far east, where the service provided by the previous Ekran had expired some time before.

The most serious problem with these first-generation Soviet communications satellites was their short working life. From the 1980s, typical American comsats operated for seven to ten years, some even longer. For the USSR, the commercial imperatives were weaker and the competition non-existent: there were few incentives to build longer-lasting comsats. It was cheaper to replace comsats with new ones when they broke down, so launches during the Soviet period were frequent. The average life of a Gorizont was 69 months, Raduga 50 months, but Ekran a miserable 25. Exceptionally, one Gorizont even operated for ten years (Gorizont 3). When finances contracted in the 1990s, efforts were made to improve the lifetimes of these satellites and these were rewarded with new averages of 75 months for the Raduga 1 series, 74 months for Gorizont and a dramatically better 72 months for Ekran M. Gorizont 31 (1996) worked more than ten years and was not maneuvered off-station until May 2006.

Ekran M in preparation

Essentially, this period saw the tail end of the early Soviet generation of communications satellites. The other Soviet communication satellite series was Luch, sent to 24-hr orbit to provide 24-hr communications for the Mir space station. Luch addressed the problem that Soviet cosmonauts could communicate with the ground only when passing over a Russian ground station or tracking ship. With Luch, the cosmonauts communicated outward to 24-hr orbit, with the signals then relayed back to the ground. With three Luch in orbit, there was always a Luch farther out, so round-the-clock communications became possible.

Only five Luch were launched, money for the series ran out and the last Luch ceased functioning in 1998. This presented difficulties during the final years of Mir, because it meant that cosmonauts could be in contact with the ground between only 10% and 20% of each orbit when over Russian territory, quite a problem during emergencies. The cosmonauts were less worried, not minding the fact that ground control could not talk to them all the time. On the International Space Station, this was less of a problem, for the ISS was able to use the TDRS system (the American equivalent of Luch) originally built for the shuttle. Where the Russians lost out badly, though, was in data return from the experiments carried out on the Russian segment. Using conventional telemetry return to the ground, the total amount of downlinked data was only 1.4 gigabytes a day, but Luch had the potential to increase the rate of return to 720 Gb/day.

A replacement series, called Luch M, was prepared, but there was no money to launch it and it ended up in a museum in St. Petersburg. NPO PM in Krasnoyarsk since proposed a smaller version called the Luch 5A, to be launched on the Soyuz Fregat rocket and the Russian federal space plan gave it a target date of 2008. There is also a proposal for a new satellite relay system called Regul, to be used by the new Soyuz version, the TMM.

During this period, there were five Molniya launches, and five launches of the old Soviet period communications satellites, Raduga, Gorizont and Ekran, bringing them to a conclusion.

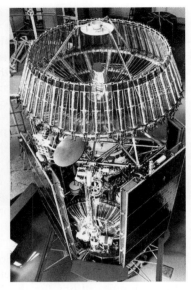

New Luch

Molniya launches, 2000–6
20 Jul   2001   Molniya 3-51
25 Oct   2001   Molniya 3-52, 3K-1
 2 Apr   2003   Molniya 1-92, 1T-28
24 Jun   2003   Molniya 3-53
18 Feb   2004   Molniya 1-93, IT-29
All on Molniya M from Plesetsk

Raduga launches, 2000–6
28 Aug 2000   Raduga 1-5 Globus
 6 Oct   2001   Raduga 1-6 Globus
27 Mar 2004   Raduga 1-7 Globus
All on Proton from Baikonour

Gorizont launches, 2000–6
22 Jun   2000   Gorizont 33
Proton from Baikonour

Ekran launch, 2000–6
7 Apr   2001   Ekran M-4
Proton from Baikonour

COMSATS: THE NEW GENERATION

Starting in 1994, Russia began to replace the old 1970s' Raduga, Gorizont and Ekran series with a new, second generation of 24-hr comsats with more power and channels and longer operating periods. Gorizont was replaced, but the attempt to replace the direct broadcast series Ekran was less successful. Here, Ekran was to be replaced by a new series called Gals: the first two were launched in 1994–95, but none subsequently.

Gorizont was replaced by Ekspress (industry code 11F639), which made its first appearance in 1994 with the launch of Ekspress 1 by Proton from Baikonour. Ekspress had 17 channels and a lifetime of at least seven years. Russia obtained allocations for six Ekspress satellites at a time—at 35°E, 42°E, 44°E, 70°E, 96°E and 335°E. The second Ekspress reached orbit in September 1996. Even at this early stage, a new version now appeared, the Ekspress A, the first of which, Ekspress A-1 was lost in a Proton failure of 27th October 1999. Happily, it was replaced by Ekspress A-2 on 12th March 2000. Ekspress A-3 followed later that year and then Ekspress A-4 in 2002.

Ekspress A-4

Ekspress AK series

The third version of the series was Ekspress AM, able to carry 38 channels, including digital TV, radio, broadband and internet and with a 12 to 15-year lifetime. With the AM series, the design company, NPO PM, made a significant effort to close the technology gap between Russian communications satellites and Western ones, bringing in the French company Alcatel. For some reason, the series started at number 22 and then number 11! Thankfully, it then reverted to Ekspress AM-1, 2, 3, etc. Ekspress AM-11 was lost after less than two years into a 12-year life on 30th March 2006 when hit by debris at 96.5°E. The satellite depressurized and lost its coolant systems immediately. Ground control maintained sufficient control to move the satellite into a disposal orbit and communications were transferred to Ekspress A-2 (103°E), AM-2 (80°E) and Ekspress AM-3 (140°E). Future Ekspress launches are planned in 2007 (Ekspress AM), 2008 (Ekspress V) and 2010 (Ekspress AT).

Ekspress was the national governmental system of communications satellites. Two other private series were introduced, one successful (Yamal), the other not (Kupon). For the first time, communications satellites were built by a company other than NPO PM. Kupon was intended to be the first of a series of new communications satellites. Launched by Proton in November 1997, Kupon 1 was the first of a three-satellite system called Bankir. Built by the Lavochkin Design Bureau, the purpose of the system was to provide satellite-based electronic banking systems between London, Russia and the Far East for the Russian Federation's Central Bank. Ideally,

Ekspress AT series

it would serve up to 40,000 terminals continuously with a satellite-delivered banking service. Unhappily for the bank, the satellite's master computer controller crashed as soon as Kupon arrived in geostationary orbit, making it inoperable, the first time such a failure had ever been recorded, though this must have been little consolation to the bankers. The rest of the series did not appear.

Yamal was built for the state gas company, Gazprom. Each box-shaped satellite weighed 1.3 tonnes and was built by RKK Energiya with American Space Systems Loral. Solar panels provided 2.2 kW of power, and 9,000 telephone circuits were installed on the satellite. Yamals were launched in pairs, the first in September 1999 (called Yamal 100 and 101, but only one deployed properly). Successors, called Yamal 200 and 201, were orbited for Gazprom by the Proton on 24th November 2003. They were designed to ensure high-quality communications between oil and gas-drilling rigs, pipeline controls and distribution channels, but there were still spare channels, which were leased to government agencies. Future Yamal launches are planned for 2007 (Yamal GK), 2008 (Yamal 200M) and 2010 (Yamal 300).

Places in the crowded 24-hr orbit were in high demand and Russia inherited 54 Soviet-period locations designated by the international agency responsible. At the end of the Soviet period, the USSR had operational satellites in 34 of these locations but these had shrunk to 24 by 2000. With the new-generation Ekspress series, Russia had even fewer satellites, but their standards had improved and their lifetimes had expanded considerably.

Despite being introduced sooner, the Molniya series was replaced later. The order had been given to develop a replacement for Molniya as far back as 1978, design being assigned to the NPO PM bureau in Krasnoyarsk and production to the Polyot company in Omsk. The first of the new series was built in 2001 and was based on the Uragan design for the navigation satellite program (see next chapter). Budget restrictions delayed the launch, which did not take place until 24th December 2006. The new Soyuz 2.1.a rocket was used from Plesetsk. The powerful Fregat stage nudged the satellite, given the name Meridian, into a low orbit. In a second burn, Fregat pointed Meridian in a $278 \times 39,801$-km, $62.8°$ orbit. Six hours later, Fregat burned a third time to raise the perigee to 1,011 km and settle Meridian into its Molniya orbit.

Ekspress launches, 2000–6

12 Mar 2000	Ekspress A-2
24 Jun 2000	Ekspress A-3
10 Jun 2002	Ekspress A-4
29 Dec 2003	Ekspress AM-22
27 Apr 2004	Ekspress AM-11
30 Oct 2004	Ekspress AM-1
29 Mar 2005	Ekspress AM-2
24 Jun 2005	Ekspress AM-3

All on Proton from Baikonour

Yamal launches, 2000–6
24 Nov 2003   Yamal 200, 201
Proton from Baikonour

Meridian launches, 2000–6
24 Dec 2006   Meridian
Soyuz 2.1.a Fregat from Plesetsk

WEATHER SATELLITES

The first Soviet meteorological satellites (metsats) had started with the Meteor 1 series in 1969, followed by the Meteor 2 series in 1974 (21 of the latter being launched when the series closed in 1993). Soviet weather satellites were a much lower priority than in the space programs of many other countries and by the end of the Soviet period no geosynchronous meteorological satellites had even been launched.

The Meteor 3 series of satellites was introduced in 1985. Five were launched during the Soviet period and only one subsequently in the 1990s (Meteor 3-6). Meteor 3 metsats weighed 2.15 tonnes (with 700 kg of instruments) and had small electric ion jets for attitude control. The 5 m tall spacecraft comprised a hermetically-sealed instrument container and two solar panels orientated to the Sun by an electromagnetic solar tracking drive providing 500-W power. Their scanning equipment was designed to observe the Earth by night as well as day. Meteor 3s carried scanning telephotometers, infrared radiometers and spectrometers, a multi-channel ultraviolet spectrometer and a radiation measurement device. They flew out of Plesetsk on 82° orbits at an altitude of 1,200 km. Meteor 3s transmitted to over 50 ground stations in Russia, using Automatic Picture Transmission—which means that foreign stations could and did use its data. Although design was originally assigned to the Yuzhnoye Design Bureau, Meteor spacecraft were subsequently transferred to another bureau with which it had close connections, the Research Institute for Electro-Mechanics, VNIIEM.

Three were normally operational at a time during the Soviet period. With the new century, three of the old weather satellites were still operating, though not all their instruments were functioning. The last Meteor 2, Meteor 2-21 (1993), was thankfully well-built and continued operating until August 2002. The last Soviet-period Meteor, Meteor 3-5 (1991), was still functioning as was the only Russian Meteor of the 1990s, Meteor 3-6 (1994).

The series was kept alive with the introduction of the Meteor 3M series (industry code 17F45), the chief designer being Rashid Salikov [1]. This was larger and heavier, 2.5 tonnes, using the Zenit launcher, with improved pointing and accuracy, able to orbit at 1,000 km altitude. The first, Meteor 3M1, carried three sets of instruments: meteorological, scientific and natural resources. These included a daytime optical scanning telephotometer, daytime and nighttime infrared radiometer, two ultraviolet spectrometers to study ozone, two scanning photometers and two microwave

Instruments for scanning the oceans

radiometers to study clouds, moisture, temperature, wind speeds and directions, ice and snow. Meteor 3M1 also carried a NASA experiment called SAGE to study ozone, aerosols, greenhouse gases and climate change.

Seven years after the previous weather satellite, Meteor 3M1 was duly launched in December 2001. The launch path brought the ascending rocket over Turkmenistan, with separation of the payload taking place over Oman, where a temporary tracking station was erected to follow progress. The Zenit 2 had spare space, so sub-satellites could be carried, being fitted to a special separation platform. These included a magnetic earthquake-predicting satellite, KOMPASS 1, the 70-kg Badr 2 (Pakistan's second satellite), a 56-kg imaging satellite called Tubsat built in Berlin for Morocco and a geodetic reflector with 32 faces (Reflektor). The last payload was—a sign of the times—flown for the United States Air Force.

The joy of at last launching a weather satellite caused the Moscow newspaper *Moskovsky Komsomolets* to gush "Poverty-stricken Russia thumbs its nose at the yankees!", proclaiming the virtues of both the micowave scanner and the microwave radiometer and how they made the Meteor better than American weather satellites. This was a dodgy claim and, in the event, Meteor 3M1 was a disappointment. There were difficulties with several of the instruments and the transmitters and the spacecraft went out of service in March 2006. An instrumentation redesign was ordered for a successor spacecraft, Meteor 3M2, which is to carry a six-channel visible and infrared radiometer, four-channel optical scanning system, 26-channel microwave sounder and imager, five-band infrared sounder, side-looking synthetic aperture radar and a sounding radiometer. The next generation, Meteor PM, is due to fly in 2012.

Despite the fact that America experimented with a 24-hr weather satellite as far back as 1966, a Soviet version was remarkably slow to appear. In 1975 the USSR announced that it would participate in the 1978–79 Global Weather Watch program by launching its own 24-hr weather satellite. This did not appear until Elektro 1 in

1994 (code 11F652), a full 16 years behind schedule. There were significant operational problems with Elektro and it ceased functioning four years later. A second Elektro was promised, but never appeared, and in the new century responsibility for weather satellites was transferred from VNIIEM to the Lavochkin Design Bureau, who promised a new version called the Elektro L, to be launched by Soyuz Fregat. Two Elektro launchings were scheduled under the Federal Russian space plan, Elektro L in 2007 and Elektro P in 2015.

Weather satellites were probably the most obvious victim of the retrenchment of the space program, with operational capacity dwindling to almost nothing. Fortunately, weather satellite systems have now been well-developed by other countries, such as the United States and Europe, and there has been a tradition from the beginning of sharing weather satellite information. Russia's dependence on foreign weather satellites attracted adverse comment in the Duma, the government responding by affirming that weather satellites would be a future priority.

Weather satellite launches, 2000–6
10 Dec 2001   Meteor 3M
              KOMPASS 1
              Badr 2
              Tubsat Maroc
              Reflektor
Zenit 2 from Baikonour

EARTH RESOURCES: RESURS DK, SICH M, MONITOR

From the mid-1960s scientists began to appreciate the ability of satellites, using different filters and light bands, to photograph the Earth's surface for the purposes of mapping, geology, agriculture, hydrology and mineralogy. The Americans flew the first dedicated Earth resources satellite, Landsat, in 1972, a remarkable spacecraft which became the basis of current space-based applications.

As with many other disciplines, the roots of Russian Earth resources work were buried deep in the Cosmos program. Cosmos was announced as the name of a new series of scientific satellites begun in March 1962, but the term "Cosmos" was used to cover a liberal range of civilian and military missions, as well as cover for unsuccessful missions. Between 1975 and 1989, 39 Cosmos satellites carried out civilian remote-sensing work, all using cabins called Zenit adapted from the photo-reconnaissance program. With *glasnost*, the Earth resources program was civilianized, taken out of the Cosmos program and given its own designation, Resurs, with a number of sub-designations. Resurs satellites were designed by the Foton bureau, part of the TsSKB/Progress complex in Samara on the Volga guided for 45 years by Dmitri Kozlov. Their signals were received by the military control center at Golitsyno, Krasnoznamensk, Moscow. Thirteen Resurs missions were flown in the 1990s and a number of replacement series was promised, with the names Resurs F2M, Resurs Spektr and Nika K, but after a while it became apparent that there was no budget to

Early Resurs series

continue the series. Resurs pictures were assembled in a huge archive which has more than 2 m photo images covering most of the world.

But, one individual Earth resources satellite was promised continually for years, the Resurs DK, the "DK" standing for designer Dmitri Kozlov. Despite having the same name as the earlier series, Resurs DK was quite different and much bigger, with a weight of 6.5 tonnes. Resurs DK was a civilianized version of the Yantar spy satellite, flying since 1974. Like the Yantar, Resurs DK was a 6 m long cone-shaped satellite with 15 m wide solar panels at its base, looking downward at the Earth. It was designed to circle the Earth for three years, taking panchromatic, multispectral and infrared pictures down to a resolution of 2.5 m (some claimed 1 m) in forty 350 km wide swathes each day, transmitting them back to Earth digitally to a number of ground stations.

Resurs DK was eventually launched on 15th June 2006 on a Soyuz U from Baikonour into an initial 191 × 338-km orbit at 70°. It immediately ran into a telemetry problem, which delayed the maneuver to raise the orbit to an operational height, but this proved to be a temporary difficulty and on orbit 49 a maneuver took place to raise the orbit to 413 × 475 km, with a view to achieving a later operational height of 600 km, which was duly achieved by the end of the month. The mission was directed by the Priroda center of the Russian Central Geodesy and Cartographic

Resurs DK

Agency and its commercial images were distributed by Sovinformsputnik. Two more satellites modeled on the Resurs DK design are due: Smotr, a gasfield monitoring satellite in 2007 and Resurs P, a cartographic and Earth resources inventory mission in 2009.

Resurs DK had some spare space and weight capacity, so two unrelated experiments were fitted. These were PAMELA and Arina. PAMELA was a joint project with Italy and stood for Payload for Anti Matter Exploration and Light nuclei Astrophysics, a precision magnetometer designed to detect high-energy particles, galactic space rays, dark matter and dark energy. Arina was an instrument designed to detect earthquakes by sensing the changes that take place in energetic particles near earthquake zones a couple of hours before the event—such an instrument had been installed on the space station Mir. Both experiments began work soon after DK reached orbit. Later, there will be a dedicated, small, earthquake-monitoring satellite called Vulkan.

The main Soviet sea observation resources program was Okean (ocean). The series was also first tested within the Cosmos program (Cosmos 1500) and was civilianized with Okean 1 in 1988. Okean was a tapering cylinder 3 m tall, 1.4 m wide at the base where two solar panels were attached, weighing 1.9 tonnes. The series was developed by OKB Yuzhnoye in the Ukraine, but flown by Russia and directed from the Yevpatoria tracking station in the Crimea. Okean carried a multizonal scanner, microwave scanner, radiometer, visual and infrared sensors, 1.5 km resolution side-looking radar and a device to receive and transmit information from buoys. Okean played an important role in tracking polar ice, both around the North Pole and the South Pole and the marine weather environment. The last Okean, the fourth, was flown by Russia in 1994.

Resurs DK in preparation

The series continued, though, as a Ukrainian program launched by Russia and tracked through the Russian tracking system, but using an entirely different type of satellite. The first of the new missions, called Sich, was launched in 1995 and functioned until 2002. The second one, called Sich 1M, was manifested from 2001 onward, but did not fly until December 2004 when it was launched on a Tsyklon 3 rocket from Plesetsk.

Although Sich was the same weight, used the same launcher (Tsyklon 3), orbit and launching base (Plesetsk) as Okean, it was quite different in design, being rectangular, box-shaped with a single solar panel. Sich 1M carried an ultrahigh-frequency scanning radiometer, the first of its kind, to capture images of land and ocean for meteorology, geology, agriculture, disaster warning, forestry and glaciology—and even the illegal growing of poppies for opiates.

Okean

It appears that the Tsyklon launcher correctly put Sich 1M in a 281 × 645-km orbit, with a view to raising the perigee and achieving the intended standard 640 × 690-km circular orbit. When the time came to do so, the engine cut out early, leaving the satellite with the low perigee. Also put in the same orbit was a 65-kg microsatellite, KS5MF2, also known as MKITS, but called Mikron for short, which carried an imaging camera. The Ukrainian Space Agency indicated that Sich 1M's own motors could be used to raise the perigee, but independent experts doubted if it had the capacity. In the new year, the Ukrainians insisted that they could still do this, although they would get only a year's operation from the satellite, instead of the planned three and that nothing could be done for the KS5MF2. The outcome of the Sich 1M mission is not known, except that the imaging system disappointed, not achieving the required 340 m resolution. In the end, Sich 1M burned up on 15th April 2006. In the pipeline, for later, are Sich 2 (a small observation satellite) and a Sich 3 radar observation satellite.

Clearly, the Earth resources program suffered badly from the retrenchment in the Russian space program in the 1990s. In response, the space program managers sought to develop much smaller satellites that could be put in orbit by less expensive rockets. Several American companies had already developed small, commercial, Earth resources imaging satellites and much could be achieved with miniaturization. Accordingly, the Monitor system was developed in the new century. Monitor was intended to provide small-resolution images that could be directly accessed by ground stations with receivers as small as 65 cm in diameter, opening the project to personal users. The contract for Monitor went to the Khrunichev design bureau, which had devised a new type of space platform for such missions, called Yakhta (yacht) and provided its own mission control. The award of the contract to Khrunichev was part of a general move to diversify satellite production and to make the contractual process more open, transparent and competitive.

A Monitor E mockup was put into orbit by Rockot from Plesetsk on 30th June 2003, accompanied by a number of microsatellites. The first operational version flew on the Rockot launcher from Plesetsk on the night of 26th August 2005 and it entered Sun-synchronous orbit of 540 km, 97.5°. The satellite weighed 650 kg and had an operational life of five years, two optical electronic cameras with 8 m and 5 m resolution and a large-memory storage system. Communications with the probe were lost on 27th August, but this proved to be a temporary hitch and caused by a problem at ground, rather than satellite level. Then, orientation of the satellite failed on 19th October and it was lost. Monitor E was intended to be the precursor to a series of annual Monitor missions, but the disappointing outcome of the first mission must have put a question mark over them. Khrunichev defended itself by saying that problems were inevitable with an entirely new system, but space managers must have pondered whether they would have been better to stick with Kozlov's bureau.

Another small imaging satellite was due for launch during this period, Kondor. Although the Strela rocket launch of December 2003 included what may have been a mockup Kondor, the real satellite was still awaited. Also due is a small Earth resources investigation satellite called Arkon 2, due 2007, as part of the federal space program, not to be confused with the military observation satellite of the same name, although it is developed by the same company, Lavochkin. Successor satellites have also been planned by Lavochkin, called Arkon 2M and Arkon Viktoria, but await government approval.

Earth resources launches, 2000–6

(30 Jun 2003	Monitor E mockup	Rockot	Plesetsk, with microsatellites)
24 Dec 2004	Sich 1M	Tsyklon 3	Plesetsk
	MKS5MF2		
26 Aug 2005	Monitor E	Rockot	Plesetsk
15 Jun 2006	Resurs DK	Soyuz U	Baikonour

MATERIALS-PROCESSING: FOTON

Soviet experiments with materials-processing dated to the Almaz and Salyut space stations in the 1970s. Later, starting with Cosmos 1645, the same space cabin as Resurs, the Zenit cabin, was adapted as an unmanned, recoverable, materials-processing laboratory called Foton (also written as Photon), able to fly 14 days at a time. This was known from 1988 as the Foton program and twelve such missions flew between then and Foton 12 in 1999.

Foton was designed by Dmitri Kozlov's TsSKB bureau in Samara. The Foton cabin weighed up to 6.4 tonnes and orbited at 62.8° in a 220 × 400-km orbit from Plesetsk. Chemical batteries provided 400 W of power. The 2.3 m diameter cabin had a payload of 600 kg in a volume of 4.5 m^3. In order to protect equipment during the final stages of the descent to Earth, solid rockets would fire under the three 27 m diameter parachutes to cushion the final fall.

Later, Foton missions were collaborative ones with the European Space Agency, mainly flying French and German experiments and the payloads were broadened into biology and related areas. Following Foton 12, the cabin was improved and called the Foton M series. There were the following main improvements:

- Increase in payload capacity from 610 kg to 660 kg.
- Lifetime extended to 16 days.
- Orbit raised to 400 km circular.
- New coolant system.
- Lithium batteries to replace silver zinc.
- Improved attitude control to improve the pureness of microgravity.

A small external module, called Biopan, was also carried [2]. ESA's share of the payload rose from 240 kg on Foton 11 to 305 kg on Foton M-1. Foton M-1 carried 44 ESA experiments altogether. The most unusual was probably STONE and this experiment related to the strange events in Antarctica several years earlier. There, American investigators found meteorites, apparently from Mars, with what appeared to be worm-like life forms in them that had survived crashing through the Earth's atmosphere and had been preserved in the rock ever since. Some people were, naturally, skeptical. Now, rock samples resembling Martian meteorites were embedded with microbiological life to see if and how they could survive reentry.

The first M-1 mission took place on 15th October 2002, carrying 650 kg of experiments from the European Space Agency and the Canadian Space Agency. In a rare failure of the Soyuz rocket, an engine exploded, the rocket crashed back to Earth and the payload was lost in the fireball.

Foton—note science module

A replacement mission was organized. This was Foton M-2 and was launched from Baikonour on 31st May 2005, the first time a Foton had launched from Baikonour. This carried 385 kg of experiments in the areas of biology, fluid physics, materials-processing, meteorites, radiation dosimetry and exobiology. Six experiments were Russian, five European and there were seven others. The mission was controlled from the ESA payload operations center, based in Esrange, Kiruna, Sweden. It made a smooth landing 16 days later and ESA specialists were on hand to retrieve the cargo. One of the most unusual experiments on Foton M-2 concerned lichens. Here, lichens (*Rhizocarpon geographicum*) were taken from 2,000 m up in the high Sierra de Gredos mountains in Spain and installed on the Biopan experiment. Investigator Rosa de la Torre found that despite 15 days in space the lichens made a recovery of more than 90% on their return to Earth, showing how lichens could live in vacuum, extreme temperatures and solar and cosmic radiation [3]. Foton M-3 was scheduled for two or three years later. Foton M-2 also carried a new, sophisticated navigation system to track its path in orbit with great precision [4].

Biology missions had originally been carried out separately in the Bion program, developed jointly with the United States. Biologists found that there was considerable value in running dedicated biology missions and that there was still much to be learned about the biological effects of the space environment which could not be determined on manned space stations alone. Typically, Bion missions orbited for between five and 22 days in 225 × 395 km orbits at 62.8° from Plesetsk. Monkeys were flown on six of the Bion missions, along with a range of other experiments, but the use of monkeys attracted protests in the United States, with activists occupying the offices of the head of NASA at one stage. When one of the monkeys on Bion 11 died in 1996, the United States withdrew from the program, which was terminated at this point, although Bion 12 had been planned. In effect, some biological experiments appeared to have been transferred to Foton M.

Later, the Russians announced plans to revive the Bion program [5]. The TsSKB Progress Design Bureau in Samara detailed plans for a more sophisticated version, Bion M, able to orbit at 450 km for up to 60 days from Baikonour. Bion M is equipped with solar panels to supply power to 800 kg of experimentation and is due to fly in 2010. Intended passengers are rats, mice, gerbils, amphibians, reptiles, crustaceans, mollusks, fish, insects, bacteria, plants and cultures.

Foton launches, 2000–6

15 Oct 2002	Foton M-1	Plesetsk (fail)
31 May 2005	Foton M-2	Baikonour
Both on Soyuz U		

SCIENCE: KORONAS, SPEKTR

Traditionally, scientific missions also operated within the Cosmos program. At the end of the Soviet period, these were civilianized and given their own titles. An example is the Interball program of magnetospheric observatories (1995–96).

Bion

The only science program to survive retrenchment was the Koronas series of solar and magnetospheric observatories, built by NPO Yuzhnoye in Dnepropetrovsk. Koronas stood, in Russian, for comprehensive orbital near-Earth observations of the active Sun and its aim was to study the Sun's internal structure (helioseismology), to investigate solar activity, predict the impact and arrival time of solar disturbances in the Earth's atmosphere and to discover the reasons for periodic discharges and eruptions from the Sun. Its specific aims were to learn more about solar eruptions—why they took place, their structure, characteristics, cycles and solar plasma emitted. The spacecraft was in the shape of a multi-sided cylinder, with four solar panels and four instrument booms. Equipment included a heliometer, coronograph, polarimeter, photometer and X-ray telescope. Koronas was part of a three-part series: Koronas I (1994), Koronas F (F for *Fyzika*) and Koronas F (F for *Foton* this time) and involved scientists from Germany, Slovakia, France, Poland, Britain and the United States. The program was managed by the Russian Lebedev Institute.

Koronas launcher, the Tsyklon

The first Koronas solar observatory, Koronas I, was launched on 2nd March 1994. Although the successor was delayed and at one stage looked unlikely to fly, Koronas F was launched on a Tsyklon 3 from Plesetsk on 31 July 2001. Weighing 2,260 kg, it was put into a 500-km circular orbit at 85.2°, 94.7 min. After accomplishing its mission, it fell back to Earth in the Indian Ocean off Kerguelen on 7th December 2005. Both were based on the standard satellite bus called AUOS developed by the Yuzhnoye Design Bureau in Dnepropetrovsk.

Koronas Foton is to study the Sun's structure and atmosphere, the solar wind, the impact of solar energy on the Earth's magnetosphere and ionosphere, and high-energy particles. The project was held up because Ukraine was unable to contribute a third AUOS bus, but supplied a Meteor 3 model satellite instead, along with one scientific instrument. Research supervisor was Yuri Kotov of the Astrophysics Institute in Moscow. In 2005 the Indian Space Research Organization joined the project, agreeing to supply a 55-kg low-energy gamma ray telescope to detect solar and galactic radiation. Launch into a 559-km, 85.2° orbit, similar to its predecessors, was set for 2007. Koronas Foton has much improved downlink relays, increasing the rate of data return ten times [6].

Besides Koronas, Russia's other scientific project was Spektr (not to be confused with the Mir module of the same name). This was a three-part Soviet-period series of large, six-tonne observatories to be launched by the Proton rocket into high orbit around the Earth. Had this three-part series gone ahead, Russian space science in the 1990s would have been world-class and well up to the standard of the great observatory series launched by NASA at the same time. But, science was the first casualty of the retrenchment of the program in the 1990. Although the Spektr program was announced in the 1980s, it made almost no progress in ten years. In 1997 it was announced that R200bn would be injected into the project in the hope of generating some momentum. European Space Agency participation was sought and the Russians also tried to involve India. The only consolation for Russia was that when a Proton rocket launched Europe's Integral observatory in 2001, Russian scientists obtained 27% of its observing time.

With the introduction of the 2006–15 space plan, work on the series resumed once more, with a view to launching Spektr R (Radioastron) in 2007, Spektr RG (Radio Gamma) in 2009 and Spektr UV (Ultra Violet) in 2010. If they go ahead as planned, Russian space science may be set for a dramatic revival [7]. The Spektr series was de-scoped, using a new smaller bus called Navigator designed by the Lavochkin Design Bureau and using either the Soyuz 2 Fregat or Zenit 2 Fregat for launcher. The 3,295-kg Spektr R will carry a 20 m diameter radio telescope dish and orbit from 500 km to as far out as 340,000 km. Its aim will be to analyze the signals from distant galaxies, with the aspiration according to Lavochkin's director Georgi Polishcuk that it might detect radio signals from distant civilizations. The 1,850-kg Spektr RG will fly from Kourou, French Guyana into an equatorial orbit of 580×660 km and will carry European Space Agency instruments to study X-rays. The project's full name is Spektr RGRL (Roentgen Gamma Rosita Lobster).

Spektr UV, Ultra Violet, also called the World Space Observatory, will have a 1.7-m mirror to gain information on the early universe, galaxies, hot stellar atmospheres and the intergalactic environment. Weighing 2,250 kg, it will orbit from 500 km to 300,000 km above the Earth and was expected to be the world's main ultraviolet telescope in orbit between the retirement of instruments on the American Hubble Space Telescope (2012) and the Space Ultra Violet Observatory (2020). Appointed as project manager was Boris Shustov, head of astronomy at the Russian Academy of Sciences, who promised that the telescope would be one of the most important space telescopes of the decade and held out the hope that it would detect and explain large quantities of dark matter. Spektr UV's elongated orbit will give it a considerable advantage over the Hubble Space Telescope, whose low orbit was determined by the launching capacity of the space shuttle.

A number of other science programs are at the conceptual stage:

- Chibis microsatellites, to study the Earth's magnetosphere and radiation belts, to be launched from the International Space Station;
- Resonans, small 320-kg drum-shaped satellites to study the Earth's magnetic field as a model for extrasolar Earthlike planets with a magnetic field, due for launch in 2012. Intended orbits are 1,800 km to 30,000 km at 63.5°;
- Kliper microsatellites (not to be confused with the spaceplane of the same name), to fly in the solar wind at a distance of 3.5 m km away;
- Intergelizond, a solar probe to approach the Sun's poles to within 42 m km;
- Strannik, a solar weather observer to fly in Earth polar orbit from 2014;
- Celesta, a star-mapping mission and Terion, a geophysics mission, both planned for 2018.

All, of course, depended on a continued improvement in the budget, but the prospects now appeared to be better than at any time since the Soviet period.

Science missions, 2000–6
31 July 2001   Koronas F
Tsyklon from Plesetsk

SMALL SATELLITES

A trend in satellite design in the 1990s was the building of ever-smaller satellites. Miniaturization made it possible for satellites to achieve more with less, making it possible for smaller rockets to be used, thereby reducing cost. Furthermore, some rockets had spare payload capacity, which could be filled by a number of quite small satellites. They could be classified into a number of sub-categories: small satellites (less than 500 kg), microsatellites (less than 100 kg), nanosatellites (less than 10 kg), picosatellites (less than 1 kg) and even femtosatellites (100 g!). Here we look at small scientific and engineering satellites.

The Mozhayets Military Space Academy in St. Petersburg began to develop small satellites both to improve small satellite design and as a training exercise for its students. The idea was developed by General Fadeyev there and with another general and a colonel, he set up the Center for Instruction in Space Technology, TsKNOT, and persuaded NPO PM to give them spare Strela 3 90-kg communications satellites. The first was launched in 1996, the second in 1997 as Zeya [8]. The series continued into the new century. Mozhayets 3 was used to train the Academy's students in satellite control systems. Mozhayets 4 was built to test laser precision geodetic instruments and the effects of the space environment on radio systems. Mozhayets 5 was a different type of small satellite begged from OKB Polyot; it was designed for long-distance optical and laser communications, but sadly failed to deploy and was stranded in the rocket's top stage.

Small satellites: Mozhayets launches, 2000–6
28 Nov 2002   Mozhayets 3
27 Sep 2003   Mozhayets 4
27 Oct 2005   Mozhayets 5
All on Cosmos 3M from Plesetsk

Launcher of small satellites, Cosmos 3M at Plesetsk

Other mall satellites, 2000–6

10 Dec 2001	KOMPASS 1	Zenit 2	Baikonour
20 Jan 2005	Tatyana	Cosmos 3M	Plesetsk
26 May 2006	KOMPASS 2	Shtil	Barents Sea

In 2004, Moscow Lomonosov State University students built, with OKB Polyot in Omsk, a small 23-kg satellite to celebrate the university's 250th anniversary and it was launched into a 1,000-km, 83° orbit with Cosmos 2414 on 20th January 2005 on a Cosmos 3M rocket out of Plesetsk. It was to study magnetic fields, storms and particles and was named Tatyana in honor of the patron saint of students. Use of the satellite was intended to be open to all students.

KOMPASS stands for, in translation, complex orbital magneto plasma autonomous small satellite, built by the IZMIRAN Research Institute and was a 80-kg satellite expected to develop techniques for forecasting earthquakes. KOMPASS 1 was, as already noted, launched as a piggyback on the Meteor 3M1 mission in 2001. By contrast, KOMPASS 2 was a dedicated mission, the only payload for its rocket. KOMPASS 2 was launched on the Shtil submarine ballistic rocket from the submerged submarine *Ekaterinberg* underneath the Barents Sea. Built by the Makeev Design Bureau, it carried an electrical field analyzer and radiowave detectors to study electromagnetic radiation, orbital plasmas and the flow of particles. Although it achieved a correct orbit, $399 \times 494 \, \text{km}$, $78.9°$, things went wrong and the satellite went out of control. Amazingly, because so few stories like this have happy outcomes, control was recovered six months later in early December.

THE UNMANNED PROGRAM: CONCLUSIONS

The unmanned program in the new century was a shadow of the high summer of the Soviet period. In the course of 2000–06, Russia launched:

- ten first-generation communication satellites (Molniya, Raduga, Gorizont, Ekran);
- eleven new-generation communications satellites (Ekspress, Yamal, Meridian);
- one weather satellite (Meteor 3M1);
- three Earth resources missions, of which only one was wholly successful (Resurs DK);
- two Foton materials-processing satellites (one failed);
- one science mission (Koronas F);
- six small satellites (Mozhayets 3-5, KOMPASS 1-2, Tatiana).

Table 3.1 summarizes Russian unmanned launches, 2000–06.

Several areas of the unmanned program have reached a very low point. Worst hit was the meteorological program, where only one weather satellite was launched and its mission was disappointing. The science program almost disappeared, with only

Table 3.1. Scientific and applications launches, 2000–06

	2000	2001	2002	2003	2004	2005	2006
Soviet period comsats	Raduga 1-5, Gorizont 33	Raduga 1-6, Molniya 3-51, 3-52 Ekran M-4		Molniya 1-92 Molniya 3-53	Raduga 1-7, Molniya 1-93		
New comsats	Ekspress A-2, A-3		Ekspress A-4	Ekspress AM-22, Yamal 200, 201	Ekspress AM-11 Ekspress AM-1	Ekspress AM-2	Meridian
Metsats		Meteor 3M1					
Earth resources				Monitor E mockup	Sich 1M	Monitor E	Resurs DK
Foton			M-1 (fail)			M-2	
Science		Koronas F					
Small satellites		KOMPASS 1	Mozhayets 2	Mozhayets 4		Mozhayets 5 Tatyana	KOMPASS 2

one launch (Koronas F) and only one other, endlessly delayed project on the books (Spektr). Other programs disappeared entirely, such as Bion, Luch and Elektro.

Having said that, progress was made. Russia completed the old Soviet-period communications satellites Gorizont, Raduga and Ekran. A new set of communications satellites was introduced, Ekspress, which promises to be the basis of meeting Russian communications needs. Small satellites were built and tested in such a way as to build up Russian expertise in the area. New, lighter designs of satellites were on the horizon and even if first tests were disappointing (e.g., Monitor, Sich 1M), they were the promise of better things to come.

REFERENCES

[1] Hendrickx, Bart: A history of Soviet/Russian meteorological satellites. *Journal of the British Interplanetary Society, Space Chronicle* series, vol. 57, no. 1, 2004.

[2] Powell, Joel: Microgravity upgrade. *Spaceflight*, vol. 43, August 2001.

[3] De la Torre, Rosa: *Lichens survive in space—Biopan experiment on Foton M-2*. Paper presented to the 57th International Astronautical Federation conference, Valencia, Spain, October 2006.

[4] Belokonov, Igor and Semkin, Nikolai: *The navigational experiments on the microgravity space platform Foton M-2*. Paper presented to the 57th International Astronautical Federation conference, Valencia, Spain, October 2006.

[5] Kazakova, A.E.; Abrashkin, V.I.; Stratilatov, N.V. and Smirnov, N.N.: *Bion M satellite— unique special purpose laboratory*. Paper presented to International Astronautical Federation conference, Valencia, Spain, October 2006.

[6] Zaitsev, Yuri: Russia to launch space observatory. *Spaceflight*, vol. 48, September 2006; *Russian space science fuels up with new ideas for Earth sciences and more. Novosti*, 7th November 2006.

[7] Zeleny, Lev: Is the golden age of Russian space science still ahead? *Novosti*, September 2006; *Russian space science fuels up with new ideas for Earth sciences and more. Novosti*, 7th November 2006.

[8] Lardier, Christian: Les satellites d'étudiants Mojaïetz. *Air & Cosmos*, #1995, 2 septembre 2005.

4

Military programs

Nothing spurred the idea of a Russian military space program as much as what happened in a taiga forest in Siberia on 20th August 1959. Woodcutters came across what they called "a strange golden globe" and when they opened it up they found that it was full of camera film. Although they didn't realize it, they had stumbled across the cabin of an American reconnaissance satellite, Discoverer, code-named Corona, the first of the series to return to Earth, although not into the hands of those intended.

Officially, Discoverer was an American scientific and engineering program to see how satellites could be brought back to Earth in advance of the manned Mercury program. The first such successful mission, which followed twelve failures, was Discoverer 13 in August 1960. As the numbers indicated, getting a spacecraft back from orbit was frustratingly difficult. Discover 6 was launched on 19th August 1959 and commanded to reenter a day later: nobody was certain what happened and all the American Air Force knew was that it had disappeared somewhere in high northern latitudes. The program was, in reality, a photo-reconnaissance one to spy from orbit on Soviet nuclear and airforce facilities, supplementing the more dangerous work of the U-2 spyplane.

The loggers wisely gave the cabin over to the police and the analysis was the subject of a memorandum by KGB chief Alexander Shelepin on 12th September 1959. On foot of this, Khrushchev ordered the development of a spacecraft able to intercept Discoverer cabins. In reality, the Soivet Union was already working on a much larger spacecraft to match Discoverer, the Zenit. Chief designer Sergei Korolev had already sketched the Zenit cabin in 1958. Although the cabin was famously adapted as the Vostok spaceship in which Yuri Gagarin circled the Earth in 1961, it was used the following year for the long-running Zenit series of photo-reconnaissance satellites. Cosmos 4 was the first Zenit in April 1962.

The Soviet military space program grew rapidly in size in the 1960s, mostly operating under the cover name of "Cosmos". The scale of the military space program is evident from the fact that the Cosmos designator reached the 2,000 mark during the Soviet period and by the new century was at the 2,369 mark. With *glasnost*, the smaller, non-military part of the Cosmos series was taken out, civilianized and given its own designators. From the period of the Russian Federation onward, only military satellites have had Cosmos designators. The term used is that such missions are launched "in the interests of the defence forces".

Within the military Cosmos program, there have always been important strands and themes. The main military programs are photo-reconnaissance, electronic intelligence, early-warning, navigation and communications, with many sub-programs. A complication is that some programs serve both military and civilian objectives—for example, the navigation satellite program.

PHOTO-RECONNAISSANCE

Discoverer and Zenit were the beginning of several generations of photo-reconnaissance satellites on both sides in the Cold War. Between them, the Soviet Union and Russia launched about 800 photo-reconnaissance satellites, presenting a mammoth task of cataloging and interpretation [1]. In the United States, these were called the KH series, while in the Soviet Union these were as follows:

Table 4.1. Soviet/Russian photo-reconnaissance satellites

Generation	Date	Series name	Sub-series
1	1962–1994	Zenit	Hector, Hermes, Heracles, Fram, Orion, Argon, Rotor, Oblik
2	1974–	Yantar	Feniks, Oktan, Terilen, Kobalt, Kometa, Neman
3	1989–2006	Orlets	Don, Yenisey
4	1997	Araks	

There is no doubt that the photo-reconnaissance program suffered badly from the decline in finances in the Russian Federation from 1992. From the mid-1960s, the Russians had always had photo-reconnaissance satellites in orbit or available for quick launch and this was especially evident whenever conflict broke out (e.g., the Middle East wars in 1967 and 1973, the Falklands in 1982 and the First Gulf War in 1991). But, by the time of the war in Kosovo in 1999, Russia had only one photo-reconnaissance satellite and one electronic intelligence satellite aloft to watch and listen to the Balkan conflict unfold. By contrast, the United States had two 15-tonne Lacrosse radar-imaging satellites, three KH-11 photo-reconnaissance satellites and three smaller imaging spacecraft. Russian coverage became very spotty. Russia had no operational photo-reconnaissance satellites orbiting from September 1996 to May 1997, nor from 15th December 1999, meaning that the new century arrived without

Zenit civilian derivative, the Resurs

such a satellite in orbit, a gap not filled until the Cosmos 2370 Yantar 4KS2 Neman launched on 3rd May 2000.

The first generation of photo-reconnaissance satellites, the Zenit, ended with Cosmos 2281 in 1994. Zenit cabins parachuted back to Earth after their missions. The principal shortcoming of Zenit was that military ground controllers had to wait until the return to Earth in order to pick up and analyze the film. This problem was partly solved by the next generation series, Yantar, the Russian word for "amber". These were much more sophisticated satellites, able to settle in a parking orbit, but swoop down to skim the upper atmosphere as they zoomed in on targets of special interest before retreating back to higher orbit. Small capsules were ejected for reentry, up to two on each mission. Pictures could then be analyzed when the main satellite was still up there and the satellite could then be ordered to concentrate on new areas. Even after their return, the main cabins were often reused, up to three times. Long before the "reusable" space shuttle was invented, the Russians were already reusing their Yantar cabins.

Yantar was commissioned by Soviet military intelligence, the GRU, in 1964 and the assignment was given to the Kozlov Design Bureau in Kyubyshev, now TsSKB of

Samara. Originally, it was thought by Western experts that just as Soyuz had followed Vostok, so Yantar had followed Zenit and was based on the Soyuz design. Not so. Yantar was an original design, a fibreglass spacecraft, cone-shaped, over 8 m tall, with a long lens at the tapering end and an overall weight of 6.6 tonnes. It had a demanding specification: the ability to orbit for at least a month (much longer than Zenit), perform up to 50 maneuvers in orbit, send down two film capsules during its mission and then itself come back to Earth itself for re-use. Power was supplied by two large solar panels at the base of the system. Yantar had an unusually powerful camera lens, with a resolution of 50 cm, sufficient to identify individual tanks, aircraft and rockets. The film was wound either into the main reentry vehicle or else into two 80-cm spherical capsules on the side of the Yantar. These descent capsules, called SpKs (*Spuskayemaya Kapsula*), had a solid-fuel retrorocket and could be parachuted down as commanded (such a system was also used on the Almaz manned orbiting space stations in the 1970s). A VHF beacon attracted the attention of the recovery forces. The main Yantar, once its mission was over, retracted its lens, burned its

Yantar civilian derivative, Resurs DK

retrorockets and descended to Russia. A small retrorocket under the parachute cushioned the final stage of the descent so as not to damage the delicate and expensive camera and optical system. The main orbital maneuvering engine weighed 375 kg, used UDMH and nitric acid and had an impulse of 1,060 kN.

Yantar was a complex satellite and took some time to design and build. The first Yantar appeared in December 1974 with Cosmos 697, but it took six test missions to achieve a fully successful profile and Yantar was not declared operational until 1977. Yantar has been through several versions: the 2K Feniks, the 4K1 Oktan and the KS1 Terilen. Three versions are in operation at present: the Kobalt, Kometa and Neman.

CLOSE-LOOK: YANTAR 4K2 KOBALT

A new Yantar version was introduced in 1982 and became operational in 1984, the Yantar 4K2 Kobalt (code number 11F695), able to stay in orbit for 60 days at a time and take a close-look at targets of particular interest. Typically, Kobalts were launched into orbits of 180×370 km, orbiting the Earth at $62.8°$ or $67.2°$ from Plesetsk. Once in orbit, they would move up and down, according to the close-look target selected and send down two SpKs before returning. Such satellites were ideal for checking on the observance of arms agreements, following conflicts in progress and watching other targets which justified military curiosity. Kobalt was a substantial program, with over 80 launched. They were built by the Arsenal Design Bureau in St. Petersburg and claim an ability to pick out detail as small as 40 cm.

The potential of Kobalt as a means of looking closely at military targets was first in evidence during the Gulf War in 1991. Three Kobalts were used to dip low over the battlefield and survey the struggle under way in the desert below—Cosmos 2108, 2124 and 2138. Kobalts have quite low orbits and were it not for maneuvers every ten days or so would burn up in a couple of weeks. Each Kobalt carries out altitude-raising firings every ten days or so, apart from special maneuvers to swoop low over particular targets.

Russia launched 17 Kobalt missions from 1992 to 2000, a much-reduced launch rate compared with the 1980s. Partly, this could be attributed to diminished funding, but also to reduced military tension requiring a lower launch rate. Most important though, the 60-day mission limit was gradually extended. First, it was brought to 65 days in 1993 and to 71 days in 1994 until 120 days became the norm. As a result, Russia was able to obtain twice the value for each Kobalt mission compared with the 1980s.

Despite funding problems, Russia was able to maintain its Kobalt program throughout the period. Originally, a new Kobalt was launched almost as soon as the previous one returned. This was no longer possible in the 1990s, but the Russians still tried to put up one Kobalt a year: Cosmos 2348, 2358 and 2365 each made 120-day missions in 1997, 1998 and 1999, respectively. The last, Cosmos 2365 returned to Earth on 15th December 1999 after 120 days, but no replacement mission was flown in 2000 at all.

Four Kobalts were flown over 2001–06. The first Kobalt in the new century was Cosmos 2377 in May 2001, which returned on 10th October 2001 in Orenburg after a 133-day mission, a record duration. Kobalt 2387 returned on 27th June 2002 after a 122-day mission which involved the standard ten orbital maneuvers. Cosmos 2410 was brought back early in January 2005 after 107 days but was never found.

A new version, Kobalt M, was introduced in 2006, also built by the Arsenal Design Bureau. Cosmos 2420 orbited for 77 days, much less than the previous standard, but appears to have been recovered normally. The commander of the space troops, General Vladimir Popovkin, announced the military's intention of launching one a year until 2010.

Kobalt launchings, 2000–6

Date	Kobalt	Duration (days)
29 May 2001	Cosmos 2377	133
25 Feb 2002	Cosmos 2387	122
24 Sep 2004	Cosmos 2410	107
3 May 2006	Cosmos 2420 (Kobalt M)	77

All on Soyuz U from Plesetsk

MAPPING: YANTAR 1KFT KOMETA

A military mapping version of Yantar was introduced in 1981, the Yantar 1KFT Kometa (code 11F660), starting with Cosmos 1246. This Yantar design actually took elements of the earlier Zenit cabin because the latter suited its instruments better. Typically, Kometas flew for 45 days out of Baikonour on 70° orbits. This series was much less prolific—once done, a mapping exercise normally lasts for some time and does not need frequent rechecking. Kometa carried two cameras—a main mapping camera taking frames of $200 \times 300 \, km$ and a precision mapping camera for frames of $180 \times 40 \, km$. Kometa flew higher than the Kobalt, at between 200 km and 270 km. Kometa used a laser altimeter to measure its precise height over the ground—essential for accurate mapping. The standard of mapping of Kometa is believed to be outstanding, providing extraordinarily precise ordnance survey types of maps. At one stage the United States Air Force bought Kometa images of Washington DC—the Air Force had been so busy photographing other people's territory that it had done no mapping of its own—and was amazed at the quality, with buildings like the Capitol, the monuments and the White House standing out absolutely clearly.

Sixteen Kometas were flown under the Soviet program from 1981 to 1991, falling to two under the Russian space program in the rest of the 1990s: Cosmos 2185, 2284 and one failure, 2243. After 2000, there were two Kometas, 2373 and 2415, so the series continues to be operational.

Kometa missions, 2000–6

Date	Kometa	Duration (days)
27 Sep 2000	Cosmos 2373	47
2 Sep 2005	Cosmos 2415	44

Both on Soyuz U from Baikonour

INSTANT INTELLIGENCE: YANTAR 4KS2 NEMAN

Kobalt and Kometa satellites, although they were able to send capsules with film down for analysis, still presented the problem that data could only be analyzed afterwards. What was needed was a satellite able to transmit data which could be interpreted in real time, before any cabin returned to Earth. The Yantar KS series was therefore devised to return digital imaging data continuously in real time. The Kozlov bureau sought to overcome the problem by returning images electronically, onboard devices scanning pictures and transmitting them back to Earth almost immediately. The Americans mastered this technology first, introducing it with the KH-11 satellite in 1976.

Digital imaging began with the Yantar Terilen series (4KS1), which orbited six to eight months and then, from 1992, with an improved version called Neman (4KS2), which orbited for ten months or so. Launch was typically from Baikonour on a Soyuz U into a nearly circular orbit of 225 km with an initial inclination of 65°. Regular engine boosts maintain this altitude quite strictly. The resolution of its cameras is thought to be in the order of 2 m. Unlike the previous models, the Yantar does not return to Earth. Instead, after a year's work, it is de-orbited to destruction over the Pacific Ocean. Nemans are built by TsSKB Progress in Samara and named after a river in northwest Russia.

Neman has the same conical shape as the previous Yantar, its long lens peeping downward toward Earth, solar panels at the top. The real break with the previous Yantars is that its two data relay antennas are directed not downward to mission control but outward toward 24-hr orbit. No signals have ever been picked up from this generation of satellites. It is thought that they beam their digital information continuously up to Potok satellites whence they are relayed back down to the ground using a relay system called Geizer (also spelt Geyser), with satellites in 24-hr orbit which send the images to ground stations in Golitsyno in Moscow and Nahodka in the far east. Thus the establishment of the Neman system also involves a commitment to maintaining a network of Potok relays in 24-hr orbit using the Proton rocket (see below). It is therefore an expensive system at a time of scarce resources.

Cosmos 2183 was the first Neman and it demonstrated its versatility by making over 40 orbital maneuvers—about five a month—before it de-orbited after a record 314 days circling the Earth. Cosmos 2223 extended the mission duration to 372 days while the next, Cosmos 2267, went even further, 419 days and flew at an angle of 70°, extending the coverage of the program to latitudes above the Arctic circle. The first in the new century was Cosmos 2370 and, two months later, there was a further launch of the Potok system of relays (2371), presumably to maintain the system for relaying the Neman's signals. Cosmos 2370 remained in orbit for exactly a year, until 3rd May 2001. There were no further launches during the period. Several times before, it has appeared that the series was coming to an end, but these rumors turned out to be premature. A satellite and mission are estimated to cost about R20m.

Yantar Neman

Date	Yantar Neman	Duration (days)
3 May 2000 *Soyuz U from Baikonour*	Cosmos 2370	365

ORLETS AND THE RIVERS: DON AND YENISEY

What is now considered to be the third generation of Soviet photo-reconnaisance satellites was introduced with Cosmos 2031 in 1989. The design code is 17F12 and it is sub-divided into two sets, the Don and the Yenisey, continuing the tradition of naming them after famous Russian rivers, the Neman being the first.

Dons followed a standard flight path of an orbit of 230×300 km, with missions being in the order of 60 to 124 days and they carried wide-angled cameras. Dons proved to be a puzzle. No pictures were released, nor hinted at, though if it is a Yantar-derived system, one may make a reasonable guess. Dons carried eight film capsules which were detached from time to time during the mission. Although it was thought that Dons could transmit information digitally as well (either directly or through the Potok system), this may not have been the case. Dons were designed by Vladimir Grodetsky of the Ishevsk Radio Plant.

Dons have the unmistakable characteristic of exploding at an altitude of 200 km at the end of their missions, normally after the eighth film capsule has been recovered. Sometimes the debris spreads far out in orbit—up to 1,100 km high—although most normally burns within a month. There is as yet no explanation why this is done. Russia may lack a means to recover them or for some other reason prefers to self-destruct them, for fear of advanced equipment falling into foreign hands. Such deliberate destruction runs counter to growing international concerns about the problem of space debris. Although previous Cosmos reconnaissance satellites carried explosive devices, explosive detonation was used only when previous missions ran into difficulty: here, explosive destruction was used as a norm. The practice appears to be to command an explosive detonation over Russian territory about an orbit after the last film capsule is sent down [2].

Don missions have proved to be infrequent. Three flew in the Soviet period and then three more in the rest of the 1990s: two in 1993 and then a long gap until 1997. None had long lifetimes, certainly not compared with the Neman series. Cosmos 2225 blew up on 18th February 1993, after 57 days in orbit, 2262 after 102 days and 2343 after 124 days. The first Don in the new century, Cosmos 2399 on 12th August 2003, was the seventh. There were numerous reports in the Russian press that the mission

Samara on the Volga, where Orlets is designed

hit snags from November onward, with fragments appearing beside the satellites. These could have been return capsules that failed to de-orbit, or, according to one story, a reel of film that unspooled in the open. In the end, Cosmos 2399 was intentionally blown up on 9th December. The next launch, Cosmos 2423, was three years later on 14th September 2006 and accompanied by reports that it would be the last and that Don systems would be replaced by a new generation of digital image transmission satellites. Cosmos 2423 blew up into 36 pieces on 17th November after 65 days, half the length on orbit as its three predecessors. The Don series was expected to be replaced by a new series called Persona in 2007 [3].

A second version of the Orlets series was introduced, the Yenisey. In May 1994, Russian television unexpectedly reported that the TsSKB Kozlov Progress works in Samara had created a new large spaceship "capable of spotting matchsticks from orbit", but that it was so secret that, even in conditions of *glasnost*, no further information could possibly be revealed. It appeared not long afterwards on 26th August 1994 with the launch of Cosmos 2290, orbited by Zenit from Baikonour, the first photo-reconnaissance satellite to take such a powerful rocket. Zenit can carry a payload of up to 13 tonnes into orbit, but that does not mean that Cosmos 2290 was necessarily that heavy (subsequently, a figure of 10.2 tonnes was given). In the event, it entered an orbit of 212×293 km, 89.5 min, $64.8°$. There are reports that it had up to 22 recoverable capsules, so it is possibly like the Don but with two bands of recoverable capsules. Data are transmitted digitally as well. The Yenisey is designed to operate for up to a year.

Cosmos 2290 made a series of engine firings to maintain its perigee at 200 km but several times raised its apogee, first to 350 km, then 450 km and finally 577 km, presumably to test its matchstick-finding abilities from such greater heights. It de-orbited over the Pacific Ocean northeast of New Zealand on 4th April 1995 after a mission of 221 days.

However, little more is known about it and the Kozlov bureau has guarded its secrets remarkably well. A second Yenisey followed in 2000, Cosmos 2372, also on a Zenit 2. Again, almost no information was given, except for a payload weight of 12 tonnes. 2372 carried out a number of maneuvers in orbit, adjusting its perigee several times to as low as 188 km and then moving the high point out to 447 km. 2372 remained in orbit 207 days, almost seven months, until 19th April 2001. Unlike the explosive Dons, the Yeniseys burn up in the Progress destruction zone in the Southern Ocean, apparently after sending down the last film capsule.

Orlets launches, 2000–6

Date	Orlets	Launcher	Type	Duration (days)
25 Sep 2000	Cosmos 2372	Zenit 2	Yenisey	207
12 Aug 2003	Cosmos 2399	Soyuz U	Don	120
14 Sep 2006	Cosmos 2423	Soyuz U	Don	65

All from Baikonour

SPACE TELESCOPE: ARAKS

An experimental photo-reconnaissance satellite was introduced in 1997, the Araks. On 6th June 1997, a Proton rocket put up a new military satellite, Cosmos 2344. After a couple of maneuvers over the following two weeks, it settled into an unusual orbit of 2 hr 10 min, 63.3°, 1,509 × 2,732 km. The Russian news services, admitting that it was a spy satellite for the Main Intelligence Directorate, the GRU, referred to it tantalizingly as "Project 11F664" and said it had been ten years in preparation. Just as some Western observers thought the Russian military space program was in terminal decline, it had sprung another surprise with the launch of the biggest spycraft ever.

The failure or the refusal of the Russians to give further details was grist to the mill of Western sleuths and spacewatchers. Startled by its orbital parameters, the indefatigable Dr. Geoff Perry, the guru behind the Kettering Grammar School space-watchers, spotted Cosmos 2344 and its Proton launcher fly over Bude later that evening and was able to recalculate its orbit, in the event even more accurately than the official announcement. Later, Dr. Perry was able to identify Cosmos 2344 by putting bits and pieces together from different news and information items given out by Russian companies over the previous number of years.

First, a search through literature given out by Russian optical companies at air shows found reference to a civilian "Project Arkon". He found a drawing from the Lavochkin Design Bureau which referred in the same context both to Arkon and Lomonosov, the latter an open project for a stellar mapping satellite. Both featured similar satellites in the shape of a telescope, with a shade on top and solar panels at the side. A Lavochkin souvenir brochure given out at the Moscow Air Show used the same illustration, but captioned it Cosmos 2344, a spacecraft capable of high-resolution photography from highly elliptical and circular orbits. Arkon appears to be a corruption of two Russian words: AR from Araks, the original name of the satellite and KON, from conversion of the satellite to civilian use ("konversiya" in Russian).

According to official reports, Araks had to be launched in 1997, otherwise its electronic systems could no longer be guaranteed. Construction of the spacecraft had begun ten years earlier, but completion had been delayed due to financial shortages. The concept was for a satellite flying 3,000 km out, with a reflecting telescope and a lens able to zoom in on military targets with a resolution of 2.5 m. The opportunity for a launch arose because the Khrunichev company was anxious to test out the new Block DM2M upper stage which would be used in a number of forthcoming launches of American payloads. The Main Intelligence Directorate, the GRU, need not have worried, for both the block D and the Araks performed perfectly and within a week the GRU analysts were reportedly "in seventh heaven" with their new toy. Araks had been expensive to develop—its R100m mission cost was five times that of a Neman—but it was expected to last three years.

Analysis of Araks suggested that it is able to take frames of $1,800 \times 3,000$ km up to an angle of $45°$. The telescope was thought to use a Cassegrain-type optical system, to be 6.89 m long and indeed to have a focal length of 27 m. The diameter of the mirror was estimated at 2 m. From its high orbit, its telescope could swing in many directions and identify objects over 1,000 km off its ground track, enabling it to spy over large areas. Ironically, after the Gulf War, the Americans identified the need for such a spy satellite themselves: their view was that a slow-moving high-flying satellite with a powerful imaging system might be of more value than some close-look satellites. The problem was that the Russians built it instead. In effect, Araks was a space telescope pointed not outward to the skies but back toward the Earth. Its more distant orbit gave it a superb vantage point for identifying, revisiting and surveiling a variety of military targets, both from directly above and from slant angles. The weight of the spacecraft was estimated to be in the order of 6.2 tonnes.

Cosmos 2344 made a big maneuver at the end of September—but information reaching Russian space enthusiasts suggest that this was not a true maneuver, but the violent depressurization of the fuel tanks. The mission appears to have terminated at this point, less than a tenth of the way into its intended flight.

A second Araks was Cosmos 2392 launched by Proton into an orbit of $1,512 \times 1,762$ km, 2 hr, $63.47°$, giving it a much lower apogee than its predecessor. It is possible that this mission also fell short of expectations, because when Stanislav Kulikov was dismissed from his post as head of the Lavochkin Design Bureau a year later in 2003, there was a long charge sheet of failed or disappointing missions, with the Araks missions included among them. Undaunted, Lavochkin returned to the fray and at the Zhukovsky Air Show in 2005 put on display models of a new, completely different, small Earth-observation satellite. The name? Arkon 2.

Araks launches, 2000–6
25 July 2002   Cosmos 2392
Proton from Baikonour

ELECTRONIC INTELLIGENCE: TSELINA

Observing the enemy by photographs is perhaps the most obvious method of spying. Equally important is observing the other side's capacity and movements through its use of radio and radar. Satellites which do this are termed ferrets or elints (electronic intelligence). Ferrets do two things: first, they listen in to enemy radio traffic and pick up conversations; and, second, they pick up enemy radar and radio frequencies and facilities, enabling their electronic equipment to be located and classified, such as anti-aircraft and ship radars. Each source has a distinctive signature according to type, size and purpose. From the information collected by ferrets, it is possible to follow enemy military maneuvers and deployments, or, as it is put in military parlance, the Electronic Order of Battle (EOB), just as effectively as visual imaging. Just as it was important for Russia not to be "blind" and without photo-reconnaissance satellites, it was equally important not to be "deaf" and without electronic intelligence. Russian ferrets are called Tselina: their payloads were originally designed and built by the Yuzhnoye Design Bureau in Dnepropetrovsk in Ukraine, with some parts later moved to the Arsenal Design Bureau in St. Petersburg.

The first generation of Soviet ferrets appeared in 1967 and was called the Tselina 1 series, with variants Tselina O and D, coming to an end in 1993. A second generation, Tselina 2 (code 11F644), was approved in 1973 as a new type of more advanced electronic intelligence satellite and construction assigned to the Yuzhnoye Design Bureau [4]. The use of the Zenit rocket enabled very large ferrets up to 9,000 kg to be flown, although they may not actually be that heavy (indeed, they may be as light as 3,250 kg). Tselina 2 is essentially an enlarged Tselina D, 4.46 m tall and 1.4 m in diameter, with windmill-type solar panels on the bottom and turnable vane panels on the middle body and an intelligence-seeking antenna on top. The data collected are relayed back both directly to ground stations and indirectly through the Geizer data relay system.

Although designed for the new Zenit rocket, Tselina 2 was instead introduced on a Proton rocket in 1984 (Cosmos 1603 and later 1656) because the Zenit was not yet ready. Cosmos 1603 maneuvered extensively in orbit, drawing attention to itself and sparking speculation in the West—correct in the event—that a new type of spacecraft had appeared. The first Zenit-launched Tselina 2 appeared with Cosmos 1697 in 1985, intended to be the start of a new four-satellite constellation. With such a constellation of four, the entire globe is covered for electronic signals once a day. In the event, the Russians have had to content themselves with a lower rate of coverage, settling for a minimum of one operational at a time.

The series was declared operational in 1988 and seven Tselina 2s flew during the Soviet period. The initial Tselina 2 program suffered badly from the early unreliability of the Zenit launcher and three were lost—in 1985, 1990 and 1991. The Russian

Zenit 2, launcher of Tselina

period also began badly with a Zenit failure on 5th February 1992. Progress with the series was again halted when a Tselina 2 was lost on a Zenit launcher which exploded 48 sec into its mission in May 1997.

Altogether, eleven Tselina 2s flew as part of the Russian space program over 1992–99. There have been two this century, the first being Cosmos 2369 in February 2000. A year later, the future of the Tselina 2 program was one of the issues discussed between President Putin during a visit to the Yushnoye Design Bureau where he met Ukrainian president, Leonid Kuchma, who had been previously the director there. Apparently, agreement was reached that there would be two more Tselina 2s to conclude the series. After two abortive countdowns, the first was launched as Cosmos 2406 in June 2004, with the second to follow in 2007.

Tselina 2s are easy to identify, all using the Zenit from Baikonour, entering very precise orbits. Orbits are very easily identifiable: for example, Cosmos 2406's orbit was 848×865 km, $71°$ and this is typical of the series. The altitude must be tight in order to obtain precise measurements of the objects surveyed below. No photographs have been published of Tselina Ds and little more is known about them or their capabilities. Assuming each has a lifetime of several years, it is reasonable to presume that at no stage were the Russians deaf during this period.

It was an interesting reflection on changed times that Russia began to give pre-launch announcements of forthcoming military missions. When the Tselina 2 planned for 31st March 2004 was delayed, the reason was even given (the need for a new control unit to be supplied from Dnepropetrovsk) as well as the revised date (25th April). An attempt was duly made to launch on 25th April, but was aborted at $T - 55$ min and on 27th April at $T - 88$ min. It was eventually launched on a Zenit 2 on 10th June 2004.

It has been announced that Tselina 2 will be replaced by Liana, built by Arsenal and TsSKB Progress, which will orbit at 500 km at $70°$. At this stage, the 1970s' design of Tselina 2 is probably quite out of date and its capabilities far below those of the current, large, American electronic intelligence-gathering satellites. The Soviet military approved, in principle, the commencement of design of what was then called the Tselina 3 system in August 1981, but little progress was made by the time the Soviet Union disintegrated. The new Liana may use either the Tsyklon 2K rocket or the larger Soyuz 2.

Tselina 2 launches, 2000–6
 3 Feb 2000   Cosmos 2369
10 Jun 2004   Cosmos 2406
Both on Zenit 2 from Baikonour

MARITIME ELECTRONIC INTELLIGENCE: US P LEGENDA

There is an important subset of ferret: the EORSAT, or Elint Ocean Reconnaissance SATellite, whose role it is to track the electronic signals of enemy fleets at sea. The Soviet Union originally devised two quite different systems to track American and

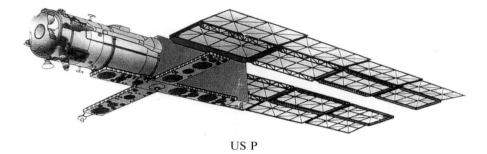

US P

NATO fleets. These were RORSATS and EORSATS. RORSATs, or the US A program ("US" being *Upravleniye Sputnik* or Russian for "controlled sputnik", with the "A" for "active"), used radars to locate and identify American ships. Radars gulped electrical power and required nuclear energy sources. After several highly-publicized reactor accidents and possibly disappointing performance, the RORSAT program was abandoned near the end of the Soviet period (1988).

The US P ("P" for "passive"; industry code of 11F120) program began with Cosmos 699 in 1974, becoming operational as the Legenda system in 1978. US Ps are built by the Arsenal Design Bureau in St. Petersburg. Thirty-six were launched in the Soviet period, ten in the Russian period in the 1990s and three this century. Originally, US P worked in pairs, which was probably adequate for routine peacetime operation. The system declined at the end of the Soviet period, to the point that by the end of March 1993 there were no US Ps operational.

Nevertheless, the system was made a priority once again and the US Ps benefited from resources transferred from the now terminated US A program. There was an upsurge of launches, with a constellation of four operating in 1994 and six by 1997. So much importance was assigned to naval surveillance that it was the only form of military space activity to actually expand in the 1990s. The Russians tried to extend their operating period from the 18 months for which it was originally designed to up to two years. The design of the US P was such that the residual propellants explode when the satellite has reached the end of its natural life. Traditionally, the Russians made a final maneuver to place the satellite in a low orbit, designed to lead the satellite to decay about two weeks later. Sometimes, though, they pushed the satellites too far and they exploded before they reached this point (e.g.. 2347, 2367).

This restoration of 1997 was not sustained for long and the system collapsed at the end of the century. Cosmos 2347 exploded in orbit in November 1999 and its companion Cosmos 2334 fell back into the atmosphere the following month. After another brief gap without a functioning US P, Cosmos 2367 was launched, making it the only operational satellite in the Legenda system as the new century came in. Cosmos 2367 blew up into 300 fragments on 21st November 2001 and there was a gap before it was replaced a month later by Cosmos 2383, the first in the new century. Cosmos 2383 was de-orbited the following year and there was a gap in coverage until Cosmos 2405 could be launched in May 2004. It was announced as a new variant, the US PM. Cosmos 2405 retired by maneuvering out of its operational orbit on 19th

Tsyklon, launcher of US P

April 2006, leaving Russia without an operational satellite for the third time. This deafness lasted for two months, replaced by a US P on 25th June, Cosmos 2421.

Typically, US Ps orbit from Baikonour at 65°, 404 × 417 km. Like Tselina 2, they must operate in precise, unchanging orbits, so as to accurately triangulate the objects they are tracking: to do so, they use ion microthrusters which frequently readjust their flight paths. Traditionally, the length of operation of US Ps in the 1980s was in the order of 200–300 days, but this reached 770 days in the 1990s.

US Ps are a 3.3-tonne main spacecraft in the shape of a long, thin torch, 17 m long and 1.3 m diameter, with two huge solar panels on either side and a large X-shaped antenna for picking up the electronic traffic. Orientation is maintained by four 10-kg UDMH and nitrogen tetroxide engines fed by eight 60 l tanks. The function of the X-antenna is to pick up the signal and identify a precise location, while ground analysts determine the type of ship involved. The launcher used is the Yuzhnoye-built Tsyklon 2, introduced 1969, later improved as the Tsyklon M and the US P program is its only use (the term Tsyklon 2 is still applied).

Just as photo-reconnaisance satellites zoomed in on ground wars, so too do US Ps closely follow fleet movements during times of conflicts or tension. EORSATs tracked the British task force as it sailed into the south Atlantic in summer 1982 to retake the Malvinas Islands during the Falklands War. Presumably, they will have followed American fleet movements in the Gulf War and in that area ever since.

How is this electronic intelligence relayed back to the military intelligence, the GRU and the Russian Navy? Like the Neman series, Tselina 2 uses a series of relay satellites. Signals picked up by the US Ps are transmitted by a series of navigation and relay satellites, Parus. Through Parus, US P information is passed on to submarines at sea, guided missile cruisers and a dedicated control center in Noginsk built specially to handle the ocean surveillance program. At least 20 Soviet naval vessels, surface and submarine, are also equipped with systems to handle incoming data from US Ps (the unfortunate submarine *Kursk* was so equipped). The aim of the system is not just to locate and identify naval ships, but to provide targeting data that can be fed directly into anti-ship missiles. Just like Neman, maintaining the US P system has required Russia to maintain another satellite program alongside.

US P production ceased in the mid-1990s, and stored satellites have been used since. The US P series is to be replaced by Liana, built by Arsenal and TsSKB Progress, which take the place of both US P and Tselina 2 and will cost R50m a go. It is unlikely that there are many Tsyklon 2 rockets left.

US P launches, 2000–6
21 Dec 2001   Cosmos 2383
28 May 2004   Cosmos 2405 (US PM)
25 Jun 2006   Cosmos 2421
All on Tsyklon 2 from Baikonour

MILITARY COMMUNICATIONS: STRELA, GONETZ, POTOK

Russia operates three military or military-derived communications systems: Strela, Gonetz and Potok. From the early 1960s, the Soviet Union defined a need to provide a state-of-the art communications system, Strela, between its military commanders, its navy on the high seas and other far-flung outposts. One of these programs, Gonetz, was subsequently civilianized. The Potok system was a series of relays in 24-hr orbit to broadcast the digital photo-reconnaisance images filmed by the Neman spy satellites.

Development of the first military communications system began in the summer of 1964 (18 August) when the USSR put three satellites into orbit on one rocket—Cosmos 38, 39 and 40. As they entered orbit, a spring was fired, pushing each of the three satellites into the required orbit in turn. As a result, they become spaced out. The idea was that a ground radio could use one, then another, then another satellite as each came over the horizon. The principle was that messages could be relayed from abroad to a Strela satellite for retransmission when passing over a control center in Russia itself several hours later. Strela is used primarily by Russian intelligence abroad, whether located in embassies or even by spies in the field and sent in code rather like text messages on a mobile phone. The system was called Strela, or arrow, to symbolize the speed with which messages could be communicated globally. The Cosmos 38–40 trio was followed by 5-in-1s and then 6-in-1s (called the Strela 3 series, still operational), 8-in-1s (called the Strela 1 series, since concluded) and single satellites (Strela 2, also concluded). The Strela series was approved in 1961, with construction originally assigned to Yuzhnoye Design Bureau in Dnepropetrovsk.

The Strela 3 series began in 1986 (Cosmos 1617–1622) and has averaged one launching a year, a rate which is maintained. Strela 3s (code 11F13) were developed by NPO PM, built by NPO Polyot in Omsk and have a guaranteed life of three years. Sixteen Strela are required for an operational system. Five sets of Strela 3 were launched in the new century, generally two at a time. Strela 3 satellites fly at 1,400 km, 114 min at 82.6°, traditionally taking a Tsyklon 3 rocket from Plesetsk. Strela are cylindrical-shaped, weigh 230 kg, 1.6 m tall and 80 cm in diameter with solar cells on the outside and a boom at either end. Inverted, cone-shaped communications aerials are attached to the back. A new version of Strela is called Rodnik. The Strela launch of October 2005 was once initially referred to as a Rodnik, and there is the suggestion that the Rodnik may actually have been introduced earlier, as far back as July 2002.

Although Strela was and is a military program, part of it was civilianized in the 1990s. This is called Gonetz, the Russian word for "messenger". Gonetz is the same as Strela, but with civilian rather than military transmitters. Gonetz was first tested out on Cosmos 2199 and 2201 in 1993, but started to use the civilian designation of Gonetz D from the next launch. From 1996 the two types were launched at the same time, with three Gonetz normally accompanying three Strela 3s. Precisely how the division of functions operates between the two was unclear. Gonetz are able to transmit text, picture, voice, paging, internet, mobile phone calls and e-mail all over the world, from equatorial to polar regions. By the new century, there were between 4,000 and 10,000 users of the civilian Gonetz system. There are three control centers for Gonetz: Moscow, Krasnoyarsk and Petropavlovsk.

A problem for the Strela series was that the last Tsyklon 3 was manufactured in 1994 and from then on old models in storage had to be used. Three Gonetz and Strela were lost at the end of 2000 when the third stage shut down prematurely: this was attributed to the lengthy period that the rocket had been in storage. The following launch in 2001 was the 120th and last of the Tsyklon 3 available for Strela. The military was obliged to transfer Strela to the less powerful Cosmos 3M. This meant that only two satellites could be launched at a time, either two Strela, or one Strela and one Gonetz.

The 21st October 2005 launch marked the introduction of the Gonetz D1M series, replacing the Gonetz D series operating since 1992. They were smaller, rectangular, box-shaped satellites of 200 kg mass, with solar panels able to provide 500 W and a lifetime of seven years. Gonetz D1M weighed 250 kg, was also built by NPO PM, had a five-year life and had a small maneuvering motor ("alignment engine"). Users included the Atomic Ministry, Arctic institutes, the Interior Ministry and the SRN (state security). It was announced that, eventually, there would be

Gonetz

Gonetz D-1M

twelve Gonetz D1Ms in four constellations of three in 1,500-km orbits, facilitating up to 100,000 users.

The other military communications system is Potok. As noted in the discussion above, the Neman series of digital photo-reconnaisance satellites required a 24-hr satellite relay system, called Geizer or Potok. This was introduced with Cosmos 1366 in 1982 and operationally in 1987 as Cosmos 1883. Each spacecraft weighed about 2.5 tonnes, was built by NPO PM in Krasnoyarsk and required the expensive Proton launcher to reach 24-hr orbit. No diagram of Potok was ever published, making the system one of the more impenetrable within the Russian intelligence-gathering system.

Potok launches are infrequent, with only one in the new century. The Potok system comprises three satellite slots, located at 190°E, 80°E and 346°E, but the Russians have rarely had more than two operational at any given time, generally moving one into position before the end of life of its predecessor. Cosmos 2319, launched 1995, was operational at the turn of the century, but it drifted off-station from 346°E in autumn 2001, by which time 2371 was aloft and went to 80°E. This may well have been sufficient for the satisfactory operation of the system. At several times, it appeared that the system had come to an end, but the system had obviously been maintained.

Gonetz, Strela 3 launches, 2000–6

28 Dec 2001	Cosmos 2384-6	Tsyklon 3
	Gonetz D1 1, 2, 3	
8 Jul 2002	Cosmos 2390	Cosmos 3M
	Cosmos 2391	
19 Aug 2003	Cosmos 2400	Cosmos 3M
	Cosmos 2401	
23 Sep 2004	Cosmos 2408	Cosmos 3M
	Cosmos 2409	
21 Oct 2005	Cosmos 2234	Cosmos 3M
	Gonetz D1M	

All from Plesetsk

Potok launches, 2000–6

4 Jul 2000	Cosmos 2371

Proton from Baikonour

NAVIGATION SATELLITES: PARUS, NADEZHDA

The USSR began its first launches of navigation satellites in 1967. Working on a similar principle to the military communications satellites, constellations of communications satellites can provide extremely accurate coordination and reference points for ships at sea, submarines underneath it and aircraft in the air.

The USSR operated a low-altitude navigation system and a high-altitude system for global positioning. All launches took place within the Cosmos label. Subsequently, the low-level system was divided into a civilian satellite system (Nadezhda, the Russian word for "hope"), and a military system called Parus, which still operates within the Cosmos label. The civilian satellites belong to what is called the Tsikada system, Russian word for a "chirping cricket". This once operated within the Cosmos program, but then acquired a designator in its own right, though only Tsikada 1 was launched.

The first generation of low-altitude military navigation satellites was phased out in 1978 and replaced by the current Parus system (Russian word for "sail"), which was originally introduced with Cosmos 700 in 1974. Parus operated in a constellation of six satellites to ensure global coverage, each satellite being replaced as it reached the end of its lifetime.

Parus comprises small satellites of about 800 kg in weight. They are cylindrical in shape, about 2 m in diameter, 2.1 m in length and equipped with a gravity boom. The navigational accuracy of Parus is estimated to be in the order of 180 m. Their main role is to provide accurate navigational fixes for the overseas submarine fleet of the Russian Navy, but they are also considered to have a key function in transmitting data from US P EORSATS both to maritime ground control in Noginsk and to ships and submarines. Granted the low level of sorties of the Russian Navy in the 1990s, the relay role of Parus may well now have become the dominant one. The priority which Russia gave to maintaining the US P system in the 1990s has of necessity required the Parus system to be kept operational as well. Parus launches continued during the 1990s, slowing toward the end. The first Parus in the new century was Cosmos 2378 in June 2001—the first since August 1999—and four were launched altogether.

Nadezhda was a civilian navigation system originally introduced within the Cosmos program. The Nadezhda system, begun under that title in 1989, consists of a constellation of four satellites each of about 810 kg orbiting at 83°, 1,000 km altitude. The series was maintained into the 1990s, with two launches in the new century, in 2000 and 2002. The M series was introduced in the 1990s, but the original numbering for the series was maintained. The differences between the first Nadezhda series and the M series are not clear, but there are reports that the M series is much smaller. The M series is due to be replaced by a new model, called Sterkh, built by OKB Polyot in Omsk. Sterkh has a weight of about 100 kg, is 1 m high, is 40 cm in diameter and certified to fly five years.

Nadezhda carries a search-and-rescue satellite system called COSPAS/SARSAT, one of the quiet and least publicized triumphs of the space age. The COSPAS/SARSAT beacon is a transponder system which ships in distress may use to summon help. It was established jointly by the Soviet Union, United States, France and Canada at the end of the 1970s. Since 1982 it has saved more than 5,500 lives on more than 1,800 rescue calls and is widely used in the West as well. Over half a million beacons are now in operation linked to the system. A ship in distress will emit signals every 50 sec to be picked up in real time by an overflying satellite. COSPAS/SARSAT is most prominently associated with the rescue of round-the-world yachtsmen and trekking climbers in the Himalayas, but most of its work is more mundane but no less critical: merchant shipping in difficulty. Because Nadezhda and the Western satellites overfly most of the world, except some of the extreme polar regions, distress signals

COSPAS/SARSAT

can be picked up within less time than one orbit. The average waiting time for receipt of a distress signal is about half that time, about 44 minutes. Generally, a Nadezhda satellite will get a fix on the distressed ship to an accuracy of less than 5 km, which should be sufficient for searching planes and ships.

Parus launchings, 2000–6

8 Jun 2001	Cosmos 2378
4 Jun 2003	Cosmos 2398
23 Jul 2004	Cosmos 2407
20 Jan 2005	Cosmos 2414

All on Cosmos 3M from Plesetsk

Nadezhda launchings, 2000–6

28 Jun 2001	Nadezhda M6
26 Sep 2002	Nadezhda M7

Both on Cosmos 3M from Plesetsk

NAVIGATION: GLONASS

In 1982, the Soviet Union began to test a high-altitude navigation system, GLONASS, which paralleled the American Global Positioning System (GPS) introduced in 1978. GLONASS stands for *Globalnaya Navigatsionnaya Sputnikovaya Sistema*, or global navigational sputnik system. The satellites themselves are called Urgan, or hurricane in Russian. GLONASS uses the Proton rocket; it is a valuable but expensive system to maintain. GLONASS is what is called a dual-use system, having both military and civilian customers. In 1998 the Russian government transferred the GLONASS system from the Defence Ministry to the Russian Space Agency as a prelude to civilianization. It has always had a separate budget line in the annual space budget.

Like the GPS, signals from the high-orbiting satellites are able to give system holders very precise estimates of their position. Originally, the accuracy was supposed to be 65 m, but in fact GLONASS provided an accuracy of about 20 m in its civil version and 10 m on its military signal (the military signal code has been quite easy to crack, as demonstrated by a hacking professor in Leeds University in England). The main users are ships in the Soviet navy and merchant navy, and military and civilian aircraft. A ship, or aircraft, must be able to receive such signals from three satellites, either simultaneously or in succession. The number of users in Russia has been small, compared with the West and were confined mainly to the military and air traffic: it was not available to personal users, hill walkers or car owners as it was in the West. > From 2003, restrictions prohibiting the use of the American GPS were lifted and many Russian enterprises used the American service immediately. Responsibility for the program falls to NPO PM in Krasnoyarsk, though they are actually built at NPO Polyot in Omsk.

GLONASS is intended to operate in a constellation of 24 satellites (21 plus three spares) in 64.8° orbits out to 19,000 km with a period of 675 minutes. Eighteen

GLONASS in preparation

satellites are necessary to cover the Russian landmass, but they are unable to provide full global coverage (only 90%). Its orbit makes it attractive for ships and aircraft operating in northerly latitudes, an important consideration for Russia, especially for the cross-polar and northern Siberian air routes. These are large satellites, 7.8 m tall and 7.2 m across their solar panels, weighing 1,260 kg each. They are cylindrical, with a signal box array at the bottom pointing down toward Earth. The GLONASS system is launched in trios on Proton rockets and reached its operational complement of 24 in March 1995.

The original GLONASS satellite had a design life of about three years, which meant that the system required fairly regular renewal. With financial shortages in the 1990s, the Russians struggled to maintain the GLONASS system. Two GLONASS launches were required each year to do this—a launch rate the Russians were unable to maintain: there were no GLONASS launches in 1997 or 1999. The system fell into decline: by 1998, only 16 satellites of the 24-satellite system were actually operating (some say only 13) and only six, the lowest point, in 2001.

President Putin took an interest in GLONASS early in his presidency and asked that the full constellation of 24 be restored by 2009. Investment was increased and it was made a defined budget line in the space program with a typical figure of over R2bn a year. After a long and damagaing three-year gap, the first set of new GLONASS was orbited at the end of 2001. Eleven were operating by 2004 and 14 by the end of 2005 [5]. For some reason, the launches came to be concentrated on the end of the year and on the Western Christmas day. The Christmas day launch at the end of 2006 brought the complement up to 17 operational satellites. These launches brought the total number of GLONASS satellites to 92, an indication of the considerable investment put into the service over the years.

In autumn 2006, Defence Minister Sergei Ivanov announced the decision to drop the military-users-only rule, offered a 30 m accuracy to civilian users and announced that GLONASS would be open to foreign customers by 2009. President Vladimir Putin expressed the view that more should be done to open up the GLONASS system, and within weeks Sergei Ivanov announced that all restrictions would be lifted at the end of the following month and that any civilian user, be that a company or an individual, could benefit from its 1-m accuracy from 1st January 2007.

An improved version, called the GLONASS M, was introduced in the 2001 launchings and four were launched by the end of 2005, though the use of the Cosmos designation remains unaffected. Thirteen were ordered. Each weighs 1,415 kg and has a lifespan of seven years, double its predecessor, which should, over time, halve the pace of replacement. GLONASS M has newer antenna and a second civilian frequency. The December 2003 and 2004 launches included two GLONASS and one GLONASS M, two M and one standard in 2005, while the 2006 launching was the first in which all were the M model. Sometimes, but not always, the M one was identified (e.g., Cosmos 2404 in the 2003 set). Two of the 2005 satellites encountered difficulties when their fuel froze, but this problem was resolved.

From 2007, a new version called the GLONASS K, with half the weight (745 kg), a 20-m accuracy and a 12-year service life, is to be introduced. It would be "lighter, smaller, cheaper, better" with a third civilian frequency: an order of 27 satellites was

GLONASS M

placed at NPO PM. Because of its smallness, new launching options were possible: either six at a time on a Proton, or two at a time on a Soyuz 2, or even an Indian launcher. It was intended to place the new satellites in orbit over 2008–10 and last to 2022, to be followed by GLONASS KM (2015–35).

GLONASS was the subject of an agreement signed in India in March 2006 by President Putin. This provided India with civilian and military access to GLONASS, giving Russia a cash injection into the system in return. The agreement reportedly provided for at least one GLONASS M to be launched on the Indian GSLV launcher.

To build support for the system, the Russian authorities decided to extend the user base of the system, aiming at a target of 80,000 users in 2007. A television program broadcast in 2006 showed how GLONASS was used, in the city of Yaroslavl, by the fire service and school buses.

GLONASS K

GLONASS launches, 2000–6

1 Dec 2001	Cosmos 2380, 2381, 2382
25 Dec 2002	Cosmos 2394, 2395, 2396
10 Dec 2003	Cosmos 2402, 2403, 2404 (M)
25 Dec 2004	Cosmos 2411, 2412, 2413
25 Dec 2005	Cosmos 2417, 2418, 2419
25 Dec 2006	Cosmos 2424 (M), 2425 (M), 2426 (M)

MILITARY EARLY WARNING SYSTEM: OKO, PROGNOZ

When Iraq launched its Scud missiles against Saudi Arabia and Israel during the Gulf War of 1991, the only possible advance notice of an impending attack was a satellite picking out the hot gas plumes of the Scud as it left its mobile launch pad. Although the Gulf War was the first to actively involve the detection and shooting down of missiles fired in anger, the use of infrared detectors to give early warning of impending missile attack was the focus of large efforts by both the United States and Soviet Union during the Cold War. The United States operated two systems—Midas, from 1960–66 and the Defence Support Programme (DSP) satellite series from 1970. The Soviet system was called Oko. It had the additional role of monitoring nuclear tests and civilian rocket launches.

The Soviet early-warning system was introduced in the 1970s and was designated the US KS system, or Oko, the Russian word for "eye", for short. It was declared operational in 1982, although the software systems continued to give trouble and in 1983 Sun reflections generated a completely false nuclear alarm. Oko operated in Molniya orbits (600 × 39,700 km, 63°, 718 minutes) which have low perigees, generally in the southern hemisphere, but high apogees, generally in the northern hemisphere (see Chapter 3). The climb to and descent from apogee is a slow, lazy one and was set in such a way as to watch out for land missiles coming from the United States, specifically the Minuteman missile. In their 12-hr orbits, Okos would swing slowly over the northern hemisphere twice a day, curving slowly to their apogees high over Asia or North America. Their telescopes could spot the ascending flame of a rising missile against the stellar background within 20 to 30 sec, enough time to alert the anti-missile forces.

Oko is drum-shaped, with a shaded telescope pointing toward Earth and two solar panels at the bottom. It is 2 m tall, has a diameter of 1.7 m and weighs 1,250 kg dry, or 2,400 kg with a full fuel load. Built by the Lavochkin bureau, its main instrument is a 350-kg, 50-cm-diameter, Earth-pointing infrared telescope to detect radiation from ascending missiles, shaded by a 4-m conical sunshield, topped by an instrument bus and two solar panels which deliver a total of 2.8 kW of power. This is supplemented by a number of smaller, wide-angle telescopes. Attitude is maintained by 16 liquid fuel engines and maneuvers can be made by four liquid fuel engines. Small orbital corrections are required every 80 days to adjust for perturbations in Earth's orbit.

Oko presented, for a military program, unusual opportunities for amateur satellite watchers [6]. First, the satellite transmits strong signals back on 2286, 2292, 2298 and 2304 MHz which could be picked up by Western radio enthusiasts. Second, the final transfer burn of the Molniya upper stage normally took place over a triangle of Uruguay, Chile and Argentina. As soon as it was over, the stage would dump any remaining propellant in order to prevent any later violent mixing of fuels and subsequent explosion. If the propellant dump took place at dusk, it would cause an expanding onion-shaped sky, many in the classic shape of flying saucers and prompting a distinct regional class of flying saucer stories. But, going back over the sightings from 1977, space commentator Jim Oberg was able to connect the pattern of unidentified flying objects to Oko burns in high orbit.

Eighty-seven Oko satellites have been launched since the series began. It is understood that Russia needs a minimum of four Oko functioning at a time to constitute an operational system, but five is preferable and nine is ideal. Russia may have often fallen short of this and the only periods when a complete constellation was fully operational was during the 1980s. The average lifetime of each Oko is about three to four years. Generally, this is a reliable, well-established system, using proven technology and the relatively inexpensive Molniya M launcher.

There is a dedicated building for Oko in Plesetsk adjacent to one of the Molniya pads there. Oko signals are transmitted to a control facility called Serpukhov 15 in Kurilovo, 70 km southwest of Moscow, near Kaluga. There are eight ground stations:

- Pechora, Azerbaijan;
- Mingetchur, Azerbaijan;
- Mukachevo, Ukraine;
- Sevastopol, Crimea;
- Gulchad, Kazakhstan;
- Irkutsk;
- Murmansk;
- Naranovichi, Belorussia [7].

Although the rate of Oko launchings fell in the 1990s, there was no indication that Russia intended to abandon the program, and it must be considered a continuing priority. A problem with Oko is that it really requires a minimum constellation to be worthwhile. If only a small number of satellites is operating, there is a danger that a potential missile launch detected by one satellite cannot be verified by another, leading to the risk of false alarms. With only a small number operating, there cannot be 24-hr coverage, which means that if Russia's enemies attack when Oko is out of coverage, the launch will not be detected. In 2001 it was announced that Russia intended to re-build the system and when Cosmos 2388 was launched in 2001, it became part of a five-satellite constellation of 2388, 2340, 2342, 2351 and 2368.

This was a good operational system once again, but it lasted for only a month. Serpukhov 15 burnt down in a 16-hr blaze in May 2001. Caused by an electrical fault, the seven-story building (three above ground, four below) burnt down and blazed all day despite the best efforts of one hundred Kaluga fire-fighters, Ministry of Defence fire crews and tankers of foam. Documentation was saved, but the center was out of action until September, which meant that Russian had no early-warning capacity from this orbit at all (indeed, until August, none from 24-hr orbit either). Without signals to maintain their orbits, three Okos (Cosmos 2340, 2342 and 2351) drifted so far out of their paths as to become useless, so this was a real blow to the system. When ground control resumed, 2368 was still operating, 2351 was recovered and there was a further launch, Cosmos 2393, later that year. But this meant that the five-strong constellation was down to three. After a long gap, the first new Oko for several years was launched in July 2006, Cosmos 2422. By this stage, only Cosmos 2388 and 2393 were still operating, so 2422 brought the system back to three satellites, far fewer than what was desirable.

A basic drawback with the Oko system was that it did not provide global coverage. Oko could detect rockets heading for northern Russia over the pole—the most likely direction from which they would come—but could detect neither launches from the continental landmass of the United States as they took off, nor submarine-launched missiles. To do so, a system based in 24-hr geosynchronous orbit was required. In 1979 the government approved the establishment of a warning system in 24-hr orbit. In 1981 the USSR requested and was allocated seven satellite slots in 24-hr orbit for Earth observations, calling the system Prognoz. This was a confusing thing to do, for there was already an unrelated satellite series of solar observatories called Prognoz. Strictly speaking, the satellites are called US KMO satellites and belong to the Prognoz system (code 71X6). An early-warning system in 24-hr orbit required a minimum of two satellites to achieve reasonable effectiveness and four to form an operational system. The first four slots were located over the priority observation areas of the United States, the Atlantic, Europe and China.

It was and is still not clear if the Prognoz system was intended to supplement or fully replace the Oko system. In the event, it has supplemented Oko, which has continued. Prognoz has only ever used four of the seven slots allocated (336°E, the main one, but also 12°E, 80°E and 35°E) and these missions have been far from trouble-free. Requiring a heavier Proton booster, it was a much more expensive system than the Oko. Not a single Proton launch of a US KMO has failed.

US KMO system

Cosmos designations were also given for this series. The first test version was Cosmos 1546 in 1984, followed by the first operational version, Cosmos 2133, in 1991. Six launches were made in the 1990s, with one in the new century. To start with, the first Prognoz slots were filled by old Oko satellites adapted for a new role in 24-hr orbit.

Cosmos 2133 was the first US KMO second-generation early-warning satellite sent to the Prognoz 24-hr orbit. US KMO is made by NPO Lavochkin, weighs 3 tonnes and has a 600-kg Cassegrain optical assembly with aluminium-coated beryllium mirrors which scan the Earth's surface every 7 sec and is so sensitive it can pick out the afterburner of a jet fighter. It has sunshades to protect it against laser attacks.

Ideally, each US KMO satellite should last five years, but the program has suffered from US KMO satellites failing long before their time. Several failed after less than a year and Cosmos 2350 lasted only two months when a seal broke in the instrument compartment. At the turn of the century, none was operational. With the Oko system out of action because of the May 2001 fire, Russia had no early-warning coverage at all until the arrival of Cosmos 2379 over 80°E in August 2001 (that October it began a move across to 336°E where it arrived in December). 80°E is ideal for watching the Far East, but 336°E (alternately expressed as 24°W, over the Atlantic) is better for watching the United States and supporting the Oko network. When Cosmos 2397 arrived in April 2003, it was sent to 80°E, but it appears to have depressurized on the way there and did not become operational. The Russians normally like, as a minimum, to have one both at 80°E and 336°E.

Following the fire, the system was so depleted that the combined Oko and Prognoz systems would take some time to recover. As a result, their benefit was marginal and could not be relied on for the detection of a surprise attack (although they could detect a massive attack). Some have questioned whether Russia needs the

US KMO system, another view

system at all, for it is based on the axiom that the United States is the main potential adversary, a proposition which has not been sustainable for many years [8]. At the end of 2006, Russia appeared to have only one operational US KMO Prognoz (Cosmos 2379) and four US KS Oko (Cosmos 2422, 2393, 2351 and 2368).

US KS Oko launches, 2000–6
  1 Apr 2002   Cosmos 2388
24 Dec 2002   Cosmos 2393
21 Jul 2006   Cosmos 2422
All on Molniya M from Plesetsk

US KMO Prognoz launches, 2000–6
23 Aug 2001   Cosmos 2379
23 Apr 2003   Cosmos 2397
Both on Proton from Baikonour

THE MILITARY SPACE PROGRAM: CONCLUSIONS

Granted the state of the Russian economy in the 1990s, it is remarkable that Russia was able to maintain a military space program at all. Unlike some of the civilian programs, military programs cannot attract in outside investment and depended entirely on the state budget. Moreover, the operation of the military space program required the use of two of the most expensive rockets in the fleet: Zenit and Proton.

The launch rate of Russian military satellites contracted sharply. Russian military launchings fell from 28 in 1992 to only five in 2000. During the March 1999 Balkan war, Russia could deploy only one photo-reconnaissance satellite and one electronic intelligence satellite, while the Americans had no fewer than eight such satellites in orbit.

A reduction in activity was likely in any case due to factors other than finance. Even if the government had not changed in 1992, there would have been a much reduced rate of launching, as satellite lifetimes were extended and frequent, but ultimately uneconomical methods of putting satellites into space were phased out. More value was obtained from existing systems by extending their lifetimes. Furthermore, the reduced level of military tension in the 1990s meant that Russia could afford to operate its military space program at a lower level. Russian and American missiles were no longer targeted on one another and few generals on either side saw the other as a direct, on-going threat. There was a significant scaling down in the nuclear arsenal of both superpowers and two neighboring countries effectively disarmed their deterrents (Ukraine and Kazakhstan). American intelligence increasingly focused on China as its main potential nuclear enemy and considered that the most likely future threats were likely to come from countries like North Korea and Iran. If one were to make a comparison with the 1980s' ocean-going navy and Soviet-supported insurgencies in Africa, Russia's own foreign policy ambitions were much diminished.

Thirty-seven military launches were made in the new century. At one stage, in autumn 2000 and for the first time in five years, Russia had three photo-reconnaissance satellites in orbit: Cosmos 2370 Neman, 2372 Orlets Yenisey and 2373 Kometa. This was exceptional, for during most of this period with some programs the Russians struggled to maintain a viable military space program. There were multiple periods of blindness (several months) and deafness when Russia could not observe or listen into naval military targets (several days) or operate a viable early-warning system. The Russians clearly struggled to maintain the GLONASS system. Within the photo-reconnaissance program, continuous coverage no longer appears to be possible. Only one Neman was flown, instead of being replaced a year at a time. Russia appears to be dependent on a series of periodic short missions by Kobalt, Don and Yenisey missions.

Periods without photo-reconnaissance (blind)
4–29 May 2001
10 Oct 2001–25 Feb 2002
27 Jun 2002–25 Jul   2002
 9 Dec 2003–24 Sep 2004
10 Jan 2005– 3 May 2006
17 Nov 2006–

Periods without maritime electronic intelligence (deaf)
23 Nov 2001–21 Dec 2001
21 Feb 2004–28 May 2004
19 Apr 2006–25 Jun 2006

Table 4.2. Russian military launches, 2000–06 in Cosmos series

	2000	2001	2002	2003	2004	2005	2006
Yantar Kobalt		2377	2387		2410		2420
Yantar Kometa	2373					2415	
Yantar Neman	2370						
Orlets Yenisey	2372						
Orlets Don				2399			2423
Araks			2392				
Tselina 2	2369				2406		
US P		2383			2405		2421
Strela/Gonetz		2384-7 D1 1,2,3	2390-1	2400-1	2408-9	2234, D1M	
Potok	2371						
Parus		2378		2398	2407	2414	
Nadezhda		M6	M7				
GLONASS		2380-2	2394-6	2402-4	2411-2	2417-9	2424-6
Oko			2388, 2393				2422
Prognoz		2379		2397			

Periods without early-warning system
May–September 2001

The program was modest, but diverse. The continuation of the digital-imaging program and the ocean surveillance system US P required the operation of two relay systems, Potok and Parus, respectively. Here, Parus was maintained, but Potok operated at a limited level. The Tselina 2 program continued to provide on-orbit electronic intelligence coverage by at least one satellite. Table 4.2 summarizes Russian military launches, 2000–06.

As can be seen, Russia launched four Kobalt, two Kometa, one Neman, one Orlets Yenisey, two Orlets Don and one Araks in the photo-reconnaissance program. In the area of electronic intelligence, Russia launched two Tselina 2 and three US P satellites. In the area of communications there were one Potok and five Strela 3/ Gonetz launches. In the navigation system there were six GLONASS sets and four Parus. In the early-warning system there were three Oko and two Prognoz satellites.

REFERENCES

[1]   Phillip S. Clark: Classes of Soviet/Russian photo-reconnaissance satellites. *Journal of the British Interplanetary Society*, vol. 54, no. 9–10, September–October 2001.
[2]   Phillip S. Clark: Final equator crossings and landing sites of CIS satellites. *Journal of the British Interplanetary Society*, vol. 55, no. 1–2, January–February 2002.
[3]   Lontratov, Konstantin and Safranov, Ivan: *Last Russian spy satellite cracked on orbit*, posted by Jim Oberg on *fpspace@friends-partners.org* on 20th November 2006.
[4]   Hendrickx, Bart: Snooping on radars—a history of Soviet/Russian global signals intelligence satellites. *Journal of the British Interplanetary Society*, *Space Chronicle* series, supplement 2, 2005.
[5]   Ionin, Andrey: Russian space programs—a critical analysis. *Moscow Defence Brief*, undated.
[6]   Grahn, Sven: Tracking Oko from Sweden, *www.svengrahn.ppe.se* (2006). For Jim Oberg's investigations of South American UFOs see *www.debunker.com/texts/giant_ufos*
[7]   Pillet, Nicolas: *Les satellites US-K Oko. http://membres.lycos.fr*
[8]   Podvig, Pavel: History and current status of the Russian early warning system. *Science and Global Security*, vol. 10, 2002.

5

Launchers and engines

Russia managed to maintain its manned, military and civilian space programs to a greater or lesser degree. But, would it be able to develop and modernize its launching potential? How did Russia develop its rocket fleet in the new century?

The old Soviet space program was remarkably economical and used only six rockets, albeit with multiple variants: the R-7 (and its many versions), the R-12 and R-14 Cosmos, the R-36 Tsyklon, the UR-500 Proton and the the Zenit–Energiya system. In the course of the 1990s, Energiya was abandoned, but Zenit retained. A range of small rockets was introduced, generally using old hardware left over from the Cold War. Existing rockets were modernized and new upper stages were introduced. An entirely new rocket was designed, the Angara, and has yet to fly. The present rocket fleet has 18 rockets (see Table 5.1).

OLD RELIABLE

More R-7 rockets have been built than any other make in history. With the COROT mission in December 2006, an astonishing 1,716 had been launched. One Western expert calculated that the factory which built its engines must have turned out a new R-7 nozzle every twelve minutes of the working day since the start of the space age! With its core stage and four strap-on stages and twenty nozzles roaring at liftoff, it is unique in the rocket world. As the years went by, the engineers simply added more and more length to the top, till the rocket reached to nearly 50 m, compared with the more modest 29 m when it started as Russia's first intercontinental ballistic missile, the R-7A, in the 1950s. The R-7 is one of the oldest rocket designs, its development being approved on 20th May 1954 and making its first flight in 1957. It has a distinctive silhouette, comprising a long, thin core stage (block A) with four similar

Table 5.1. The Russian rocket fleet

	Stages	Length (m)	Launch weight (kg)	Payload (kg)
Proton	4	56.1	690,000	20,600
Proton M	4	57.2	723,943	5,500 (GTO)
Soyuz (U version)	3	49.5	309,000	7,500
Molniya M	4	45.2	305,000	1,600 (GTO)
Tsyklon 2/M	2	40.5	183,254	3,583
Tsyklon 3	3	49	188,000	4,100
Tsyklon 4	3	40.5	196,198	5,300
Volna	2	14	40,000	110
Shtil	3	14.8	39,916	160
Rockot	3	29	107,000	1,850
START 1	4	22.7	46,902	700
Dnepr	3	34.3	210,800	4,200
Cosmos 3M	2	31.4	109,000	1,780
Zenit 2	2	57	459,950	13,700
Zenit 3SL (Sea)	3	59.6	472,600	5,896 (GTO)
Zenit 3SLB (Land)	3	58.65	468,000	3,600 (GTO)
Strela	2	29.2	105,000	1,600
Angara (3 version)	3	49	464,000	4,500 (GTO)

GTO – Geostationary Transfer Orbit

stages clustered around it (blocks B, V, G, D). Several minutes into the mission, blocks B, V, G and D peel off, leaving the long, thin block A to continue to fire until the third stage is ready to take over.

There have been no fewer than 16 versions of the R-7. These were given the traditional set of project codes, starting with the first test before Sputnik (8K71) to the last Soyuz version, the U2 (11A511U2). Some have been fairly minor variants, with only one or two launches. The most used versions have been the Voskhod model (11A57, 306 launches) which was used for all but the very last phase of the Zenit photo-reconnaisance program, Molniya M (8K78M) (with Molniya, 318 launches) and the Soyuz U (11A511U) (878 launches by December 2006). The Soyuz U was a consolidation of a number of 1960s' variants and the U design has remained unchanged since. It was intended to replace the U with the more powerful U2, but the plant making its fuel additive, sintine, closed, so it was taken out of service after 70 launches. So, the Soyuz U soldiers on.

In the early 1980s the Progress factory at Samara was making R-7s at the rate of 50 a year, but this was scaled dramatically back to fewer than twelve a year. With the investment of the Western company Starsem and its success in winning Western contracts, the rate of construction was increased to over 20 a year by the new century. The introduction of a number of new, successful upper stages to the Soyuz and the new launch site in French Guyana will likely, if anything, lead to further production runs.

In a program under financial pressure, the R-7 offered huge advantages, for the

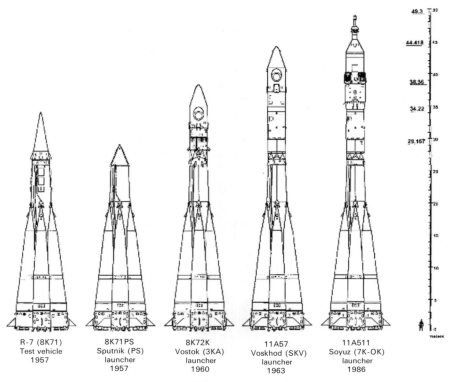

R-7 (8K71) Test vehicle 1957	8K71PS Sputnik (PS) launcher 1957	8K72K Vostok (3KA) launcher 1960	11A57 Voskhod (SKV) launcher 1963	11A511 Soyuz (7K-OK) launcher 1986

R-7 evolution

main development costs and problems had been paid for and sorted out thirty years earlier. For example, the original Molniya version was introduced for the first Mars probes in 1960. More than forty years later, it was still perfectly suited for the role of putting Molniya communications satellites and Oko early-warning satellites into high orbits over the northern hemisphere.

R-7 failures are infrequent. There were two in the mid-1990s, only a month apart, when a new launch shroud was introduced without a full appreciation of the consequences for the rocket during the most stressful period of ascent. The first failure of the R-7 in the new century was Foton M-1 on 15th October 2002 at Plesetsk. The Soyuz U version lost an engine on one of the side stages just 29 sec into the mission and the control computer commanded a shutdown of the whole system. The rocket fell back to Earth and exploded on impact, making a huge fireball, blowing out windows 3 km away and causing widespread damage, including the death of a soldier on guard duty. The launch of the manned Soyuz rocket then planned for two weeks later was delayed while the investigation reported. A state commission was appointed to investigate and found, in its report a week later, that a foreign object, made of iron and chrome according to its traces, had lodged in a hydrogen peroxide pipe in the engine's gas generator.

Molniya M cutaway

The second was in the Molniya M version. Here, the failure to launch Molniya 3K in June 2005 was attributed to excessive vibration in a second-stage motor. This was in turn traced to a manufacturing fault in the factory, which should have been identified when it was checked before launch.

NEW UPPER STAGES: IKAR, FREGAT

Despite its age, Russia invested considerable energy in improving the capabilities of the R-7. This focused in the short term on new upper stages and in the long term on new versions. Two upper-stage upgrades were made: Ikar and Fregat.

The first upgrade was a small upper stage called Ikar (Russian), Ikare (French) or Icarus (in English). Ikar already existed: a throwaway remark by one of the designers suggested that it had already been used in the Yantar program. Ikar, as a civilianized version, offered a perfect method for getting groups of communications satellites into 1,400-km orbit. Soyuz–Ikar won the competition for part of the American Globalstar network of low-Earth orbiting comsats, partly because of the R-7's known reliability and the price on offer, less than $40m, compared with the American equivalent, the Delta 2, which would have cost nearly $60m. Winning the competition provided the additional resources and investment necessary to redevelop the Ikar as an operational upper stage for the Soyuz.

Ikar was 2.9 m long, 2.72 m in diameter, weighed 3.29 tonnes, with 900 kg of UDMH and nitrogen tetroxide fuel and able to get 3,300 kg into high-Earth orbit. The motor was a 17D61 able to generate 2,943 kN and was equipped with 16 steering thrusters. Ikar was adapted with a 390-kg dispenser to spring each of four satellites into their appropriate orbits. All six Ikar Globalstar launches went off with perfect precision in 1999.

Fregat was the next and more durable upper-stage development. Fregat was a new upper stage introduced in 2000 to facilitate the placing in orbit of the European Space Agency's Cluster satellites to study the Sun. The original Cluster series had been lost when the first Ariane 5 exploded in a giant fireball over Kourou, French Guyana on its maiden flight in 1996. Now the European Space Agency turned to Russia for help with putting a set of backup models into their appropriate orbits. The agency required an upper stage with considerable thrust and versatility. Fregat offered the possibility of several restarts and could put five-tonne payloads into precise orbits as high as 450 km.

Like Ikar, Fregat had historical antecedents—back in the Phobos program in 1988 where it had been developed as a universal stage for Venus, Mars and Moon missions. Fregat had eight tanks carrying 5,440 kg of UDMH fuel, 28 attitude control thrusters and a rocket motor which could be used for up to 20 course corrections. Made by Lavochkin, it was 1.5 m tall, 3.35 m in diameter and could burn with a thrust of 19.6 kN for up to 877 sec. It offered the perfect solution to the European Space Agency's problem of how to orbit its Cluster series of satellites. The ESA contract provided the extra resources that enabled Fregat to be redeveloped as an operational system.

In preparation for the Cluster launch, two demonstration tests were carried out. The first Soyuz Fregat was duly launched on 9th February 2000 and carried out its two demonstration burns, one at 200 km, the other at 600 km. A month later, on 20th March, Fregat went through its paces again, with repeated engine restarts and a mock separation of the payload.

Cluster was eventually launched from Baikonour on 16th July 2000. By this stage the Cluster satellites had received names, Rumba, Salsa, Samba and Tango to reflect the way the satellites would dance in formation around the heavens. Fregat fired 90 min into the mission, putting the first two Cluster satellites into a parking orbit of 240 × 18,000 km, the first of six firings to send them to a final operational altitude of 19,000 × 119,000 km. The second set went into orbit on 9th August. The Soyuz third stage was not properly filled with propellants and underperformed, but the Fregat was powerful enough to make good the difference.

Since then the Fregat has been used seven times, not only on the original Soyuz rocket but on the subsequent FG and 2 models. Soyuz FG Fregat was used for the two European Space Agency interplanetary missions, Mars Express and Venus Express, on the Israeli Amos satellite, the Galaxy 14 commercial satellite and on the first of the new series of European navigation satellites, Giove 1. As a small innovation in the space program, it has proved remarkably successful and has been sketched in for virtually all the high-orbit and deep-space missions for the federal space plan for 2006–15. Lavochkin has since proposed an upgrade of Fregat, the Fregat SB, its weight rising from 7.3 tonnes to 12.25 tonnes, with up to 10.8 tonnes of fuel.

The next phase of upper-stage development was called the ST. This was not a stage as such, but a new fairing. The Europeans had put much emphasis into the importance of fairing design, seeing it as the key to making small but important improvements in performance. The ST shroud was derived from the European Ariane 4 rocket. Its lightness and shape meant an improved performance going up through the atmosphere, increasing the payload at 450-km orbit from 5 tonnes to 5.5 tonnes. This was introduced with the Soyuz 2 rocket (next).

RUS PROGRAM

In the 1990s the Russians began a program to upgrade the Soyuz launcher. The program was called "Rus". This was funded by the government to pay the design bureaus to find ways of improving performance. The program was managed by TsSKB Progress, where it is made.

The first upgrade was called the Soyuz FG (11A511FG), and this made its first flight on the Progress M1-6 on 20th May 2001. The features of the FG are:

- improved performance of putting 8.2 tonnes to orbit, compared with 7.3 tonnes for the Soyuz U;
- improved RD-107 and RD-108 engines, giving 5% more thrust;
- advanced fuel injector systems.

The Soyuz FG was originally intended for the manned Soyuz TMA spacecraft, which first flew in October 2002. To test its reliability, Soyuz FG was first used on four unmanned Progress launches (Progress M1-6 to Progress M1-9).

The second upgrade was the Soyuz 2. This has two sub-variants, the 2.1.a and the 2.1.b.

The Soyuz 2 (14A14) series has improved engines, a new shroud (the ST) and is able to launch 300 kg more payload (7.4 tonnes to low-Earth orbit). Soyuz 2.1.a has modernized first- and second-stage engines, digital controls and improved pumps and piping. Gone was the old-fashioned countdown, supervised by 40 people at twelve workstations in the bunker near the rocket and the old-fashioned turning of the launch key: instead, it was all done by computer. The third stage is the RD-0110 built by the KBKhA Design Bureau in Voronezh, which provides 29.8 tonnes of thrust for 240 sec. Weight is 311.7 tonnes. The new version has much-improved telemetry, permits a ground-based abort and has vertical integration of the upper stage.

The first Soyuz 2.1.a launch took place on 8th November 2004 from Plesetsk. The payload was an old Oblik satellite that had never flown and whose guarantee

Soyuz FG cutaway

Soyuz 2 first launch from Plesetsk

had run out. There is some dispute as to whether this was an orbital or sub-orbital mission, but it appears to have been successful. The third, block I, stage fired for eight minutes, about what might have been expected; its payload still attached, it impacted with the Pacific Ocean.

The Soyuz 2.1.b has a RD-0124 third stage, also from KBKhA Voronezh, with more thrust (30 tonnes) and significantly longer burn time, 300 sec, with a specific impulse of 359 sec and 900 kg more payload than the 2.1.a version. The shroud is made by TsSKB Progress and is 4.11 m in diameter and 11.43 m long. Performance is 8.6 tonnes to low-Earth orbit. The 2.1.b is the basis for the Soyuz ST B—the version which will fly from Kourou, French Guyana—where performance is 5.5 tonnes to geosynchronous transfer orbit (see Chapter 6). The RD-0124 is, in effect, the first, new third stage on the R-7 since the 1960s.

The first operational launch of the Soyuz 2.1.a was planned for July 2006, the payload being the European weather satellite Metop. This was a sophisticated satellite, Europe's first polar-orbiting satellite, carrying an impressive suite of French-built and American-derived instruments, so it was doubly important all went well. Unlike most Soyuz launches which normally went like clockwork, launching Metop proved to be a frustrating experience. The launch campaign was an exasperating one. The first countdown, on 17th July 2006, was halted with only seconds to go, with a similar experience the next day. On 20th July, during the third attempt, the launch was pulled with 3 min 5 sec to go and it was decided to unfuel the rocket and try again later. Some of the older hands watching probably shook their heads and would have preferred the old launch key and pre-digital systems. After the third disappointment, the system was disassembled so that the problem could be rooted out. When Metop was brought back to the pad on 17th October, the Soyuz 2.1 counted down in a black-dark Baikonour night, a stiff wind blowing the liquid oxygen off the side of the rocket. When the rocket counted down, the clock reached zero, but once again nothing happened, the software aborting the sequence at $T - 1$ sec. The culprit, it turned out, was a faulty umbilical cable. A fresh attempt was ordered for the next day at the same time (the Sun-synchronous orbit gave

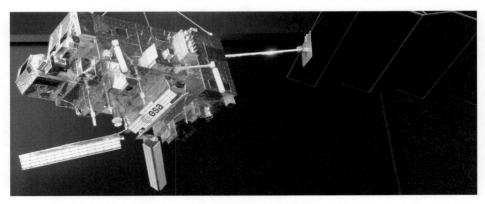

First Soyuz 2 operational launch—Metop

the same launch time each day), but high-atmosphere winds forced another postponement.

Finally, the sixth attempt was made on 19th October. This time everything went remarkably smoothly, the Soyuz 2.1.a rising on a pillar of orange–yellow flame, lighting up the low but light cloud deck as it punched through. The Soyuz 2.1.a curved over to the north, a trajectory rarely followed from Baikonour, the diamond shape of the burning engines receding into the far distance. Sixty-nine minutes later the Fregat engine released the 4,093 kg satellite over the remote French Indian Ocean island tracking station of Kerguelen, spot on-course for its 98.7° polar orbit. Control was soon taken over by the European control center in Darmstadt, Germany. Thus, on its 1,714th launch a new version of the R-7 became operational.

The next 2.1.a also proved to be a reluctant flier. On its next mission—the Meridian satellite out of Plesetsk—there was a computer problem compounded by high winds, so the launch was delayed a day. On the second countdown, there was a problem with power supplies and the countdown computer froze. It was third time lucky though and Soyuz 2.1.a put the new Meridian satellite into a high Molniya orbit.

The 2.1.b model made its first launching at year's end, orbiting the European planet-finding telescope COROT, a French-led European project. The nighttime take-off went perfectly and 50 minutes later COROT was in orbit on a three-year mission, equipped with a 30-cm telescope so sensitive that it could detect the acoustic signatures of planets transiting across the face of distant stars.

Some other future versions of the R-7 have been proposed. These are the Onega, the Yamal, the Aurora and the Soyuz 3.

Onega was a proposed Energiya–Volga branch upgrade of the R-7, designed to bring its payload up to something in the order of 11–12 tonnes. The company proposed to:

- widen the core stage from 2.06 m to 2.66 m, enabling 50 tonnes more fuel to be carried;
- widen the third stage from 2.66 m to 3.44 m, with an extra five tonnes of fuel;
- use the NK-33 rocket engine from the old Moon program for the core stage;
- for the strap-on stages, use the RD-107A engine, now used for the Soyuz FG;
- for the third stage, use the RD-0124E, the engine used on the Soyuz 2;
- for higher orbits, fit a new upper stage—a fourth stage—by developing the Molniya block L (Taimyr) or the Proton block D (Korvet).

The Volga branch later promoted the Onega for the Kliper space shuttle (see Chapter 8), proposing the following set of further changes:

- for the first stage, the RD-191 used for the Angara when it became available;
- for the strap-ons, the RD-120.10F engine, based on the RD-170 used for the Zenit;
- for the third stage, four RD-140E, a liquid-hydrogen engine developed by the KBKhA in Voronezh;

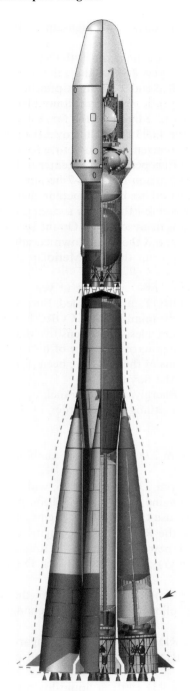

Soyuz 2 cutaway

- for the fourth stage, a hydrogen-fueled engine called the Yastreb, also developed by the KBKhA.

Yamal was the next proposal. This was, in effect, the four-stage Korvet version of the Onega. The Volga branch called it the Yamal, suggesting that it would be a suitable launcher for the forthcoming series of Yamal communications satellites. Their flattery did not win out and the Yamal comsats were eventually launched by the traditional Proton. The Volga designers then suggested an export version of Yamal, called Aurora, which was associated with plans by a Russian–Australian consortium to develop a launch base on Christmas Island [1].

The federal Russian space plan for 2006–15 envisages further upgrades to the R-7, called the Rus-M program and funding has been set aside for this purpose. The aim is to work toward a version able to lift up to 15 tonnes; this may be called the Soyuz 3 rocket. The Soyuz 3 appears to be much the same as the second iteration of the Onega, but without the Angara RD-191 engines, namely:

- for the first stage, the NK-33 engine from the Moon program;
- for the strap-ons, the RD-120.10F engine, based on the RD-170 used for the Zenit;
- for the third stage, four RD-140E engines—liquid-hydrogen engines developed by the KBKhA in Voronezh.

At this stage, it is impossible to judge whether the Onega, Yamal, Aurora or Soyuz 3 versions will fly. The relentless upgrading of the old R-7 rocket suggests that the chances are good. Chief designer Sergei Korolev's 1954 rocket design lives on.

In the course of 2000–6, Russia launched 38 Soyuz U, nine Molniya M, four Soyuz Fregat, thirteen Soyuz FG, five Soyuz FG Fregat and four Soyuz 2.

Soyuz U launches, 2000–6

1 Feb 2000	Progress M1-1
6 Apr 2000	Soyuz TM-30
26 Apr 2000	Progress M1-2
3 May 2000	Cosmos 2370 Yantar Neman
6 Aug 2000	Progress M1-3
29 Sep 2000	Cosmos 2373 Kometa
17 Oct 2000	Progress M-43
31 Oct 2000	Soyuz TM-31
16 Nov 2000	Progress M1-4
24 Jan 2001	Progress M1-5
26 Feb 2001	Progress M-44
28 Apr 2001	Soyuz TM-32
21 Aug 2001	Progress M-45
15 Sep 2001	Pirs
21 Oct 2001	Soyuz TM-33
25 Apr 2002	Soyuz TM-34
16 Jun 2002	Progress M-46

Soyuz U launches, 2000–6 (*cont.*)

15 Oct 2002	Foton M-1 (fail)
2 Feb 2003	Progress M-47
8 Jun 2003	Progress M1-10
12 Aug 2003	Cosmos 2399 Don
29 Jan 2004	Progress M1-11
25 May 2004	Progress M-49
11 Aug 2004	Progress M-50
23 Dec 2004	Progress M-51
28 Feb 2005	Progress M-52
31 May 2005	Foton M-2
16 Jun 2005	Progress M-53
2 Sep 2005	Cosmos 2415 Kometa
8 Sep 2005	Progress M-54
21 Dec 2005	Progress M-55
24 Apr 2005	Progress M-56
28 Jun 2005	Progress M-57
14 Sep 2006	Cosmos 2423 Don
23 Oct 2006	Progress M-58

From Baikonour

29 May 2001	Cosmos 2377 Kobalt
25 Feb 2002	Cosmos 2387 Kobalt
24 Sep 2004	Cosmos 2410 Kobalt

From Plesetsk

Molniya M launches, 2000–6

20 Jul 2001	Molniya 3-51
25 Oct 2001	Molniya 3-52, 3K1
1 Apr 2002	Cosmos 2388 Oko
24 Dec 2002	Cosmos 2393 Oko
2 Apr 2003	Molniya 1-93, 1T-28
19 Jun 2003	Molniya 3-53
18 Feb 2004	Molniya 1-93, IT-29
21 Jun 2005	Molniya 3K (fail)
21 Jul 2006	Cosmos 2422 Oko

All from Plesetsk

Soyuz U with Fregat stage, 2000–6

8 Feb 2000	Dumsat/IRDT 1
20 Mar 2000	Dumsat
16 Jul 2000	Cluster 1
9 Aug 2000	Cluster 2

All from Baikonour

Soyuz FG launches, 2000–6

20 May 2001	Progress M1-6
26 Nov 2001	Progress M1-7
21 Mar 2002	Progress M1-8
25 Sep 2002	Progress M1-9
30 Oct 2002	Soyuz TMA-1
26 Apr 2003	Soyuz TMA-2
18 Oct 2003	Soyuz TMA-3
19 Apr 2004	Soyuz TMA-4
14 Oct 2004	Soyuz TMA-5
15 Apr 2005	Soyuz TMA-6
1 Oct 2005	Soyuz TMA-7
30 Mar 2006	Soyuz TMA-8
18 Sep 2006	Soyuz TMA-9

All from Baikonour

Soyuz FG Fregat launches, 2000–6

2 Jun 2003	Mars Express/Beagle 2
27 Dec 2003	Amos 2
13 Aug 2005	Galaxy 14
9 Nov 2005	Venus Express
28 Dec 2005	Giove 1

All from Baikonour

Soyuz 2 launches, 2000–6

8 Nov 2004	Oblik (sub-orbit)	Plesetsk	2.1.a
19 Oct 2006	Metop A	Baikonour	2.1.a Fregat
24 Dec 2006	Meridian	Plesetsk	2.1.a Fregat
27 Dec 2006	COROT	Baikonour	2.1.b

COSMOS 3M

The Cosmos 3M was the smallest launcher from the end of the Soviet period. The Cosmos 3M started life as the Cosmos 1 (1964–65), then the Cosmos 3 (1966–68) and ever since then its improved version, the 3M. This rocket was approved on 2nd July 1958 and developed by the Mikhail Yangel bureau in Dnepropetrovsk, Ukraine, over 1958–61 (code 11K65) to launch satellites in between the 500-kg capacity of the small Cosmos rocket (R-12) and the 7.5 tonnes of the Soyuz. Essentially, it was a fatter, bigger and more sophisticated version of the long, thin R-12. Approval to develop the R-14 was given by the government on 31st October 1961.

The R-14 Cosmos 3M is 31.4 m tall, has two stages and can orbit payloads of up to 1,780 kg. On-the-pad weight is 109 tonnes, of which most is fuel. The first stage has two RD-216 engines, each weighing 1,350 kg, 2.2 m tall, with a thrust of 1,728 kN, specific impulse of 291 sec, a chamber pressure of 73.6 atmospheres and a running time of 146 sec. They burn UDMH and nitric acid.

Cosmos 3M production line

Its first mission was the launch of three Strela military communications satellites in August 1964. The Cosmos 1 version made eight launches (1964–65), the Cosmos 3 four (1966–68) until the Cosmos 3M version came in as the 11K65M in 1967, much the most successful version. During the peak years of the Soviet program, the Cosmos 3M may well have been used 25 times or more each year. The R-14 was originally produced in Krasnoyarsk (the Cosmos 3), but since then the only production line was in Omsk (the Cosmos 3M, where it was made by Polyot Enterprises). There are three Cosmos 3M pads at Plesetsk (131—a double pad—and 132) an unused one at Baikonour (41) and one at Kapustin Yar (107). Although Cosmos 1 and 3 were launched from Baikonour, the launcher was transferred to Plesetsk with the Cosmos 3M in 1967. A small number of launches, including sub-orbital tests, has been made from Kapustin Yar. The nine-in-one launch on 27th October 2005 was the 436th Cosmos 3M.

The Cosmos 3M is used as the launcher for the Parus series of navigation satellites, as the launcher of the Strela 3 series of military communications satellites following the end of the Tsyklon 3 production line and for a miscellany of multiple

RD-216 engine

missions. Production of Cosmos 3M in Omsk ceased in 1994 and for the following ten years reserve and military rockets were used. There have been reports that production resumed in 2005. In the course of 2000–6, there were fifteen successful Cosmos 3M launches.

Generally, it is regarded as a successful and reliable launcher, with only one failure during this period (Quickbird, 21st November 2000). Quickbird was a 930-kg, private, American, imaging, mapping and Earth resources mission. No signals were ever received from the satellite. The upper stage, with the satellite, broke up and crashed into the atmosphere of Uruguay the following day. A Russian investigation showed an interruption of telemetry at 400 sec while the rocket was still climbing, and this was later attributed to a deterioration in the quality of engine, which had been thirteen years in storage before launch. The Cosmos 3M has a reliability rate over its

Cosmos 3M interconnector

service career of 94%, almost all accidents being in the early stage of its development, including a launch disaster at Plesetsk on 26th June 1973 causing nine fatalities.

The Cosmos 3M was marketed in the West, the agency being Assured Space Access Inc., the rate being $10m for a full launch and less than $10,000 for small piggyback payloads, a number of which were carried (below). A modernized version of the Cosmos, called Vzliot (code 11K55, or Cosmos 3MU), was once reported in design, but did not appear to progress.

Cosmos 3M launches, 2000–6

28 Jun	2000	Nadezhda M-6
		Tsinghua
		SNAP 1
15 Jul	2000	Champ
		Mita
		Rubin 1
21 Nov 2000		Quickbird 2 (fail)
8 Jun	2001	Cosmos 2378 Parus
28 May 2002		Cosmos 2389 Parus
8 Jul	2002	Cosmos 2390, 2391 Strela
26 Sep	2002	Nadezhda M-7
28 Nov 2002		Mozhayets 3
		Alsat 1
		Rubin 2
4 Jun	2003	Cosmos 2398 Parus
19 Aug 2003		Cosmos 2400, 2401 Strela
27 Sep	2003	Mozhayets 4
		Laretz
		Bilsat
		Nigeriasat
		UK-DMC
		KAISTSat 4
		Rubin 4
22 Jul	2004	Cosmos 2407 Parus
23 Sep	2004	Cosmos 2408, 2409 Strela
20 Jan	2005	Cosmos 2414 Parus
		Tatyana
27 Oct	2005	Mozhayets 5
		Sinah 1
		China DMC
		Topsat
		SSETI Express
		Ncube 2
		UWE 1
		XI-V
		Rubin 5
19 Dec 2006		SAR Lupe 1

All from Plesetsk

PROTON AND PROTON M

The UR-500K Proton dates to 29th April 1962 when Soviet leader Nikita Khrushchev asked designer Vladimir Chelomei to build a large rocket able to deliver a huge load of themonuclear bombs on an enemy city, what he called a "city-buster". Thankfully, the Proton was never used for this ghastly purpose and was adapted as a heavy-lift rocket for more peaceful purposes, making a triumphant first flight in July 1965. Proton never had the extensive series of upgrades that characterized the development of the Soyuz, but a new version was introduced in 2001, the Proton M.

Proton

Proton's RD-253 engine

The 44 m tall Proton became in the course of time the workhorse of the deep-space program, geostationary satellites and the orbiter of heavy payloads into Earth orbit. Despite its unreliability in its early years—and problems which cost the Russians the Moon race—it became very solid and reliable, rarely malfunctioning after 1970. Proton was able to lift up to 21 tonnes into low-Earth orbit (e.g., the space station modules Zarya and Zvezda) or 6 tonnes to the Moon or Mars or 2.5 tonnes to geosynchronous orbit (e.g., Ekspress) or for the global-positioning system (GLONASS). Proton was also used for a number of domestic military programs (e.g., Raduga, Potok, Prognoz). The Proton rocket is manufactured in the Khruni-chev plant in Moscow, part of the American–Russian company International Launch Services.

Proton exists in several versions: the original two-stage (UR-500 Proton); the three-stage and the four-stage (called the UR-500K Proton K, with the fourth stage

Proton taking off

normally carrying the block D fourth stage made by the Energiya Corporation); and the new Proton M (with the Briz M fourth stage). Proton's history had a long struggle over who would build the fourth stage. Originally, Chelomei designed his own upper stage, but in December 1965, shortly before his death, Korolev wrested control of the upper stage from Chelomei and used his own bureau's fourth stage, the block D. In 2000, long after they were both dead, the Proton started flying with another fourth stage, the Briz—built by the Khrunichev company, long associated with Chelomei.

From 1983, the USSR began to offer the Proton to Western commercial companies on an economic basis. Proton eventually won its first commercial contract in 1992 with a deal to launch an Inmarsat communications satellite. Since then, a stream of commercial contracts followed to put comsats into 24-hr orbit, mainly for Western communications companies. Proton was able to offer reliability, a proven track record and costs below the American rate, enough to make a difference in the competitive global world of the 1990s. By the late 1990s, the annual launch Proton rate had, despite the contraction of the Russian economy, actually increased and as many as fourteen were launched in 2000.

Proton launchings have two unusual features. First, because the Proton uses nitric acid storable fuels, there are none of the telltale wisps of liquid oxygen burning off during the final stages of the countdown. There is no means of telling that the rocket is fueled and ready to fly—it just goes, the nitric acid evident in its telltale orange–brown smoke. Second, the bottom part looks as if it holds six strap-on rockets, like the four of R-7. In fact, they are not boosters, but fuel tanks positioned on the side feeding into the main propulsion area. Like all Russian rockets, Protons are transported on the railways. Proton is a large rocket, the bottom stage filling the width of the railway gauge: the fuel tanks must be transported separately and attached later.

Proton's reliability rate was lower than the Soyuz. Two crashed out of the sky in the 1990s, only months apart in July and October 1999. There was plenty of forensic evidence to examine and within two days investigators had found the wreckage of the engines responsible. The investigators found a pattern of poor workmanship in a batch of engines produced in the Voronezh plant in 1993 when there had been a break in production due to a shortage of orders. In the words of the investigation, "a large number of particles had found their way into parts of the engines. Ducts and welds had not been properly cleaned." Rigorous checks were run on all engines to ensure that none of the bad batch was installed on a Proton awaiting launch. It was decided that future Protons would be fitted with filters to the inlets of the gas generators, the number of rivets and welds would be reduced and the turbopumps would be made of tougher alloys able to withstand higher temperatures.

It was not just the bottom part of the Proton that caused trouble. The problem at the top end, the block D and its successor the DM, was worse. A squat cylinder, block D weighed 33 tonnes, including casing, fuel tanks, fuel, motors and instrument unit. Block D went back to the 1960s and had originally been built for the Soviet manned Moon landing but had been used as the top stage of the Proton—most numerously to get satellites to 24-hr orbit. Proton suffered four upper-stage block

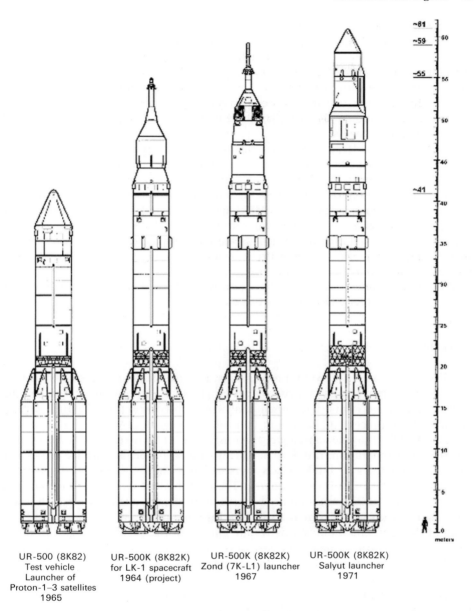

UR-500 (8K82)	UR-500K (8K82K)	UR-500K (8K82K)	UR-500K (8K82K)
Test vehicle	for LK-1 spacecraft	Zond (7K-L1) launcher	Salyut launcher
Launcher of	1964 (project)	1967	1971
Proton-1–3 satellites			
1965			

Proton evolution

D failures in the 1990s: Gorizont (May 1993), Raduga 33 (February 1996), Mars 8 (November 1996) and Asiasat 3 (December 1997).

The persistent failures of the block D must have raised questions among the Proton managers. These mishaps re-emphasized the desirability of introducing a new, more powerful and reliable upper stage. Here, the new upper stage called Briz M made its appearance. Briz M was a new upper stage, made by Khrunichev, which can be restarted up to 20 times and was intended to fly on a more powerful version of the Proton, the Proton M. It was in the shape of a flattened doughnut, 4 m in diameter but only 2.65 m tall, with the engine, the central unit and instrumentation surrounded by a toroidal fuel tank. The engine provided a slow-burning capacity to put up to 6.6 tonnes into geostationary transfer orbit. Briz M was 2.3 tonnes empty, but 22.1 tonnes when fully loaded with its cargo of nitric acid and UDMH. The engine was the S598M or 14D30 of KB KhimMash. Briz M can make multiple burns— four are standard, six are possible—and some quite long ones, for example 37.5 min on the AMC-15 launch in 2004.

What had been intended as the first flight of the Briz M, but using as a testbed the Proton K, took place on 5th July 1999. This was the first of the 1999 lower-stage failures, so whether it would have worked was never ascertained, for the Proton crashed before the engine could be lit. This was an inauspicious start to the attempt to replace a problem engine! The first Briz was tested successfully on the Proton K on Gorizont 33 and then commercially on AMC-9. By then, the new version of Proton, the Proton M, was ready and it was specifically built to fly the Briz M, which would, hopefully, put the block D problems behind.

PROTON M

In 2001 the Proton was joined by a newer, more powerful version, the Proton M, with an improved computerized control system taken from Zenit, a larger 4-m shroud, stronger inter-stage joints, lighter materials, the full consumption of fuel without residuals, more precise guidance and control, the new Briz M upper stage and a smaller debris field should something go wrong. Engines were required to burn up all their fuel after separation, rather than let any unused propellant fall back to Earth (hitherto about 12 kg of UDMH was unused). New engines raised the thrust level from 151 tonnes to 160 tonnes per engine. These improvements were intended to give it a performance of 22 tonnes into low-Earth orbit, or, the mission for which it was most intended, up to 4.9 tonnes to geosynchronous orbit. Its first launch was an Ekran M domestic communications satellite on 7th April 2001. It was successful first time. From 2001, Russia flew a mixture of old Protons, with block D, alongside the new Proton M and its new Briz M upper stage.

Despite corrective measures, the block D jinx had still not run its course. Astra 1K was launched on a Proton K block D in November 2002. Astra 1K was, at that moment, the largest communications satellite ever built, weighing over 5 tonnes, more than 8 m tall, with 40 m wide solar panels and no fewer than 54 transponders. When the time came for the block D to fire over the Ivory Coast on the first orbit, nothing happened. As was the case with an almost identical failure on the Mars 8 mission in 1996, the payload was pre-programmed to separate and deploy its solar panels, in effect believing that it was well on its way to its target orbit. The failure left the expensive satellite stranded in a 140-km orbit from which it crashed out over the

Proton M

Pacific Ocean two weeks later, costing its insurers $292m and causing the Astra Company some difficulty in meeting its communications commitments. International Launch Services finally lost their patience with Energiya's block D upper stage at this point and announced that all future launches would use the new Proton M rocket with the Briz M upper stage made by the Khrunichev company. No more block D.

Briz M worked perfectly at first, making ten successful missions on the Proton M. Then came Arabsat, also called Badr, in March 2006. The Briz apparently failed

27 min into its second of four planned burns (31 min), leaving Arabsat in a low and useless orbit. Briz M had performed perfectly on 13 previous occasions. Arabsat was taken out of orbit over the south Pacific Ocean on 24th March. The Proton M was grounded for the summer, while the cause was tracked down—a small piece of debris in the oxidizer supply. The Proton M Briz M eventually returned to service in August with the launch of the Hot Bird 8 communications satellite. In November a replacement Arabsat was eventually put into orbit (Arabsat 4B, also known as Badr 4B). The Proton first put Badr in a low, 173-km circular orbit. Briz now went into action, burning four times and bringing it up to geosynchronous orbit.

A problem for the Russians was that they still had many block Ds in stock. Although they would no longer use them for commercial, Western satellites on the Proton, there was no reason they should not continue to use them for domestic communications satellites and for other parts of the domestic program (e.g., GLONASS), and this is what they did. The block D continued in use on the Zenit rocket.

From the 1980s the Proton made about 8–10 launches a year, with 2000 being its record (14). Its overall service reliability record is 91%, including its difficult development period. The nature of the payload changed during the Russian period, with fewer launches for the national space program and more for international commercial payloads. Altogether, Proton in all its versions made 54 launches between 2000 and 2006, with a series total of 323.

Briz

Proton launches, 2000–6
Proton K
12 Jul 2000   Zvezda

Proton K with block D
12 Feb 2000   Garuda
12 Mar 2000   Ekspress A-2
17 Apr 2000   Sesat
24 Jun 2000   Ekspress A-3
 1 Jul   2000   Sirius 1
 4 Jul   2000   Cosmos 2371 Potok
28 Aug 2000   Raduga 1-5
 5 Sep 2000   Sirius 2
 2 Oct 2000   GE-1A
13 Oct 2000   Cosmos 2374-6 GLONASS
22 Oct 2000   GE-6
24 Aug 2001   Cosmos 2379 Prognoz
 6 Oct   2001   Raduga 1-6
30 Nov 2000   Sirius 3
 1 Dec 2001   Cosmos 2380-2 GLONASS
15 May 2001   Panamsat 10
16 Jun 2001   Astra 2C
30 Mar 2002   Intelsat 9
 8 May 2002   Direct TV-5
10 Jun 2002   Ekspress A-4
25 Jul   2002   Cosmos 2392 Araks
22 Aug 2002   Echostar 8
17 Oct 2002   Integral
25 Nov 2002   Astra 1K (fail)
25 Dec 2002   Cosmos 2394-6 GLONASS
24 Apr 2003   Cosmos 2397 Prognoz
24 Nov 2003   Yamal 200, 201
28 Dec 2003   Ekspress AM-22
27 Mar 2004   Raduga 1-7
26 Apr 2004   Ekspress AM-11
29 Oct 2004   Ekspress AM-1
26 Dec 2004   Cosmos 2411-3 GLONASS
29 Mar 2005   Ekspress AM-2
24 Jun 2005   Ekspress AM-3
25 Dec 2005   Cosmos 2417-9
17 Jun 2006   Kazsat
26 Dec 2006   Cosmos 2424-6

Proton K with Briz M
22 Jun 2000   Gorizont 33
 7 Jun 2003   AMC-9
10 Dec 2003   Cosmos 2402-4 (GLONASS)

Proton M with Briz M

7 Apr 2001	Ekran M4
30 Dec 2002	Nimiq 2
16 Mar 2004	Eutelsat W3A
18 Jun 2004	Intelsat 10
4 Aug 2004	Amazonas
14 Oct 2004	AMC-15
3 Feb 2005	AMC-12
22 May 2005	Direct TV-8
9 Sep 2005	Anik F1R
29 Dec 2005	AMC-23
1 Mar 2006	Arabsat 4A (fail)
4 Aug 2006	Hot Bird 8
8 Nov 2006	Arabsat 4B
12 Dec 2006	Measat 3

All from Baikonour

TSYKLON

The Tsyklon rocket had its roots in military rockets in the 1960s and in the UR-200K design of Vladimir Chelomei. Go-ahead for the Tsyklon program was given in 1965 as an intermediate booster able to lift military payloads larger than the Cosmos 3M rocket but smaller than Soyuz, aiming at a capacity in the order of 3.5 tonnes (Tsyklon was called project 11K67). It was introduced as the Tsyklon 2 or M version in 1969 as a two-stage rocket to fly the US P EORSATs and other military satellite payloads. The first iteration, the Tsyklon 3 (11K68), with a third, powerful upper stage, flew in 1977 (there was no Tsyklon 1). To add to the confusion, Soviet sources would refer to the Tsyklon 2 as the M and the Tsyklon 3 as the plain "Tsyklon". Essentially, the difference was that the Tsyklon 2, with two stages, was used for US P from Baikonour only, while the Tsyklon 3, with three stages, was used for other missions such as Gonetz, Strela 3, Koronas and Sich. Tsyklon's first two stages both burn UDMH fuel with either nitric acid or nitrogen tetroxide as oxidizer. The Tsyklon 2 could put three tonnes into orbit from Baikonour. The longer, 39-m tall, three-stage Tsyklon-3 could put smaller payloads into more versatile 150 × 10,000 km orbits from Plesetsk. There were two automated pads at Plesetsk which can launch Tsyklons rapidly in all weathers, from +50°C to −45°C. Tsyklons were built by the Yuzhnoye factory in the Ukraine, which then assumed responsibility for further development.

There was one failure, in December 2001. This was the 118th launch of the Tsyklon 3 and it seems that the third-stage engine cut off at 367 sec. The rocket reached an altitude of 190 km before falling back with its six satellites (three Gonetz, three Strela) into the Arctic east of Wrangel Island. A control, rather than mechanical fault was blamed, again attributed to the long period that the rocket had been kept in storage beforehand.

The Tsyklon concluded production in 1994. Outstanding military versions were handed over to the space forces and this enabled missions to take place for another

Tsyklon in factory

seven years. The Tsyklon 3 version ran out first. Its Sich 1M launch of December 2004, in which it underperformed, was its 120th and last launch. The Tsyklon 2 which put the Cosmos 2421 US PU into orbit in June 2006 was its 105th launch. According to NPO Yuzhnoye, only six are left in storage.

Despite the retirement of the Tsyklon 3 and imminent retirement of the Tsyklon 2/M, Yuzhnoye has attempted to keep the Tsyklon rocket going. Tsyklon was offered to Western customers, with the price for a launch estimated to be in the order of $20m. There were no takers, probably because other launchers suited their needs better. Not discouraged, Yuzhnoye announced a new version, called the Tsyklon 2K, capable of putting 2,000 kg into Sun-synchronous orbit and equipped with an apogee propulsion module called the APM 600. In connection with the joint development of Alcântara launch base with Brazil, the company began to develop a new rocket called the Tsyklon 4. This is essentially the old Tsyklon 3, but with a new third-stage motor, the RD-861K and bigger payload shroud. So, the Tsyklon story may not be over yet.

For the future, the Yuzhnoye design bureau plans the Mayak series (the Russian word for "beacon"), a family of five launchers capable of putting payloads of

Tsyklon launch

between 1,000 kg (Mayak 32) and 10,000 kg (Mayak 22) into orbit. The Mayak series will use the more environmentally-friendly liquid oxygen and kerosene. First launch was set for 2010. The bureau has also undertaken studies of a number of less conventional rockets: the microspace launch vehicle (to be dropped from a high-flying MiG-25 jet) and an even smaller launcher to fire into orbit from the top of the trajectory of a shell fired from a 203-mm howitzer (gun launch) [2].

Tsyklon 2/M launches, 2000–6
21 Dec 2001   Cosmos 2383 US P
28 May 2004   Cosmos 2405 US PM
25 Jun 2006   Cosmos 2421 US P
All from Baikonour

Tsyklon 3 launches, 2000–6
31 Jul 2001   Koronas F
27 Dec 2000   Three Gonetz, Strela (fail)
28 Dec 2001   Cosmos 2384-6 Strela 3
              Gonetz D1 1,2,3
24 Dec 2004   Sich 1M, KS5MF2
All from Plesetsk

ZENIT

Zenit originates from Mikhail Yangel's bureau in Ukraine, NPO Yuzhnoye, as project 11K77 [3]. Following what are called the 1973 *Poisk* studies, in which the Defence Ministry examined the Soviet Union's future launching requirements, Zenit was conceived as a powerful, medium launcher to lift military payloads. A new generation of electronic intelligence satellites was in planning at the time, the Tselina 2 (see Chapter 4) and the Zenit was very much considered in this context. Because of the ecological damage done by the toxic fuels of the Proton, the decision was also taken to use kerosene as fuel, with liquid oxygen. The role of the Zenit would be to fly military payloads from Plesetsk.

Zenit 2 launch

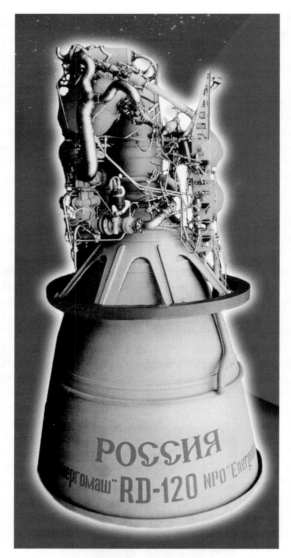

RD-120 engine

In the following year, after a meeting between chief designer Valentin Glushko and Yuzhnoye director Vladimir Utkin, the Zenit was adapted as the first stage of the huge Energiya rocket and the Buran shuttle. Part of their thinking was that the Zenit could be proven before the larger rocket flew, thereby improving its chances of success. The deal included an arrangement whereby the Zenit would have its own launch pad in Baikonour. Zenit duly flew in 1985 from there and was declared operational in March 1987.

Zenit used a rocket engine then in development by Valentin Glushko's rocket engine company Energomash, the RD-170, except that on the Zenit it was called the RD-171. The second-stage engine was the RD-120 liquid-oxygen and kerosene engine, later replaced in June 2003 by an improved version, the RD-120M. An improved version of the first stage, the RD-171M, will later be brought in and by summer 2003 had already been tested for a total of 5,500 sec. Zenit 2 has a capacity of up to 13.7 tonnes to low-Earth orbit. Its height is 57 m, weight 459 tonnes, diameter 3.7 m and its main RD-171 engine burns for 133 sec, the second stage for 303 sec.

Zenit duly served as the strap-on booster for the Energiya rocket, but, according to the Ukrainian designers' original intentions, became a powerful rocket in its own right, standing almost as tall as Chelomei's famous Proton. Zenit exists in three versions: Zenit 1 is the first stage of the now-canceled Energiya rocket, Zenit 2 is the main model and Zenit 3 is the sea-launched version used to reach 24-hr orbit (also called Zenit 3SL).

The Zenit complex was completed at Baikonour in 1985: it comprises supply and service sheds and a six-floor multi-story building capable of processing several Zenit rockets at the same time. Zenit was developed as a launcher of military satellites, with the capacity, in a situation of tension, of putting many satellites into orbit with short countdowns. Accordingly, its launch procedure is automated, fast and efficient. As soon as the payload is settled in the top of the rocket, the Zenit is transported on a railcar to its pad and put into position for firing. Once the railcar arrives, automatic systems connect the Zenit to 25 fuel lines and 3,500 electrical circuits. Zenit is clamped to the pad by vices sufficient to hold the rocket down at full thrust and then raised into firing position. Once secured, it can be fueled within minutes by pumps drawing off silos which are located underground close to the pad. The entire fueling and launch sequence is conducted automatically over a couple of hours. The hoses and pumps are withdrawn 12 min before launch. At 4 min before launch the railcar makes its way back to the integration building. A system of sprinklers is available to extinguish fires. Fifteen seconds before launch, the cooling system is activated, drowning the base of the pad in water to cool it and dampen vibration. The pad can be ready for another Zenit launch in five hours. As part of the automation process, the Zenit was equipped with a fast, high-performance computer called Bisser, originally developed for the N-1 Moon rocket by the Pilyugin bureau.

A total of 33 launch attempts had been made by summer 2000. Zenit's early and primary payload was the heavy electronic intelligence (elint) satellite Tselina 2 sent into 850-km orbits at 71° from Baikonour. By 2000, Zenit had successfully put ten Tselina 2 elints into orbit and was indispensable to the Russian electronic intelligence system. Although Zenit fired perfectly on both its Energiya flights, the rocket had a checkered history and there were numerous failures, the last in 1998. The Russians have tended to blame a fall in quality control in Ukraine, where the rockets are made and have increasingly looked to an indigenous Russian rocket for a replacement, despite the Zenit's impressive power.

Zenit 2 was used four times in the new century: twice for its original design purpose, the Tselina 2; once for the Orlets Yenisey photo-reconnaissance satellite; and once for the Meteor 3M1 weather satellite.

Zenit 2 launches, 2000–6

3 Feb 2000	Cosmos 2369 Tselina 2
25 Sep 2000	Cosmos 2372 Orlets Yenisey
10 Dec 2001	Meteor 3M1
	KOMPASS 1
	Badr
	Tubsat Maroc
	Reflektor
10 Jun 2004	Cosmos 2406 Tselina 2

All from Baikonour

UKRAINIAN ROCKETS TO THE PACIFIC: ZENIT 3SL, THE SEA LAUNCH

Zenit had an exciting new lease of life in an entirely new and unexpected venture—the Sea Launch project. This was a joint venture between the American Boeing company (40%), Russia's Energiya (25%), Norway's Kvaerner (20%) and Ukraine's Yuzhnoye (15%), to put communications satellites and similar payloads into orbit

Zenit 3SL

from a mobile equatorial launch site in the middle of the Pacific Ocean using a maritime version of Zenit.

The idea of using the Zenit from an ocean-based platform originated in studies by the company that built cosmodromes, the KBTM, in 1981. Baikonour lies at around 50°N, far north of the Equator, requiring all satellites heading for equatorial orbit to make an energy-expensive dogleg maneuver to get there. An equatorial launch site cuts out the need for such a maneuver. But, if a fixed equatorial launch site was not possible, what about a mobile, sea-based one? Sea platforms had been used to fire rockets before, for Scout rockets had been fired from the American–Italian San Marco platform off the Kenyan coast from 1967 to 1988. However, this was the first project to use an ocean-based equatorial launch platform for commercial operations to 24-hr orbit and was on a scale much larger than anything previously contemplated.

Hitherto, Zenit 2's capability to reach 24-hr orbit was very limited (only 600 kg). However, if fitted with Proton's block D upper stage and launched from near the equator, it could place 5.8 tonnes into geosynchronous orbit, making it a highly competitive launcher. Moreover, Zenit's automated countdown procedures made it ideally suited for operations in confined and difficult spaces, such as on board ship. Accordingly, it was given the new name of Zenit 3SL (3 for three stages, SL for Sea Launch).

The Zenit project involved the construction of a 34,000-tonne rocket transport, assembly and command ship built in Govan, Scotland, in the course of 1996, called the *Sea Commander*, later adapted with mission control and space-tracking gear in the Kanonersky works in St. Petersburg. This was a big ship, which rides high in the water like a modern cruise liner. The ship was 203 m long and 32 m wide, just narrow enough to fit through the Panama Canal. Inside, it comprised a huge rocket assembly hall, mission control and facilities for handling the fuels used on the launches. In the body of the ship were placed the Zenit rockets, up to three at a time, in the space that would normally be occupied by cars in a sea ferry. The Zenits filled the center of the ship in a hangar 70 m long, 12 m deep in three bays like in a railway marshaling yard. For its first journey in 1998, the *Sea Commander* traveled from St. Petersburg to Long Beach with two Zenits on board and a crew of 240.

Home port was Long Beach, California, where Boeing took over an old naval causeway which had been used as a depot from 1944, at that time a waste-dumping ground and in a terrible state. The area was cleared and rebuilt in 1997 with a hangar designed to take three Zenits and an integration building. The role of the *Sea Commander* was to load the Zenits, bring them 1,600 km across the Pacific to the launch site, transfer them to the launch platform, back away to a distance of 5 km and act as the launch control and tracking ship.

The floating launch platform was a decommissioned, 131-m-long, 78-m-wide North Sea oilrig called the *Odyssey*, bought in Norway. It was self-propelled, semi-submersible, built in Norway in the 1980s, but taken out of service when it was damaged by fire in 1988. It lay idle in Vyborg for a number of years. Bought by the Sea Launch project in 1995, it was refitted in Stavanger and Vyborg where its legs were strengthened, new power systems and a crane were installed and it was adapted

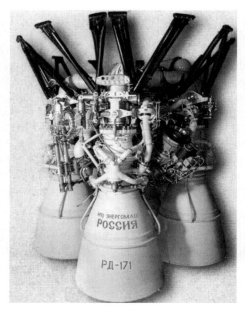

RD-171 engine

to take rockets. *Odyssey* rose 70 m above the sea. There was a considerable amount of plumbing involved in fitting the ship to take oxidizer and fuel tanks, as well as the associated loading and pressurization systems. The helipad and crew quarters for 68 were modernized and a flame trench was fitted to take away the exhausts of the RD-171 engines.

Odyssey was refitted in the Baltic and then towed out to the Pacific equatorial launch site south of Hawaii, the nearest land lying due west, Kiribati, otherwise known as Christmas Island. Getting it there was no easy undertaking. Averaging 12 knots, aided by its own four 3-m propellers, it was towed all the way from Vyborg, near St. Petersburg, Russia, to Christiansand, Norway. It barely got under the new Øresund Bridge, then being constructed to link Denmark and Sweden across the Baltic. Odyssey then proceeded down the English Channel, through the Straits of Gibraltar and the Suez Canal, across the Indian Ocean to Singapore and then across the full Pacific before reaching its home port in Long Beach. The journey took 107 days. Once it arrived there, it took on 15,000 tonnes of water ballast so it could semi-submerge and stabilize itself in the ocean. The journey from Long Beach to the Pacific site took a final 12 days.

As a general rule, rockets are transported from the the Yuzhnoye factory in the Ukraine by ship through the Black Sea, the Mediterranean, the mid-Atlantic and the Panama Canal and then up the west coast. The fuel is shipped from Russia (due to concerns about American kerosene). When the *Sea Commander* arrives alongside in mid-Pacific, it hoists the 472-tonne Zenit onto the platform, the *Odyssey*, where it is raised to the vertical for launching. For liftoff, the roof of the hangar is rolled back

Odyssey platform

and Zenit is raised hydraulically into position (the hangar then closes again). Four hours before liftoff, when fueling begins, the crew of the platform evacuates and boards the *Sea Commander*. The entire process from there on is entirely automated, as at Baikonour. The rocket stands the equivalent of a four-storey house. Launches can take place in conditions of up to 12-m/sec windspeed and wave heights up to 3 m. During the launch, telemetry is monitored from the *Sea Commander*'s mission control rooms staffed by Energiya and Yuzhnoye personnel.

The first Sea Launch mission was a demonstration mission with a mock payload in March 1999. Sea Launch soared aloft from its Christmas Island sea platform. Arcing over the blue equatorial Pacific, the block DM stage duly fired 34 min later and the payload arrived on station, over the equator, three hours later.

The first operational mission was duly made on 10th October 1999. There was almost no damage to the launch platform *Odyssey* as the Zenit soared on a pillar of flame skyward. At 2 min 50 sec, the first stage dropped off. The second stage then burned until eight minutes into the mission. There was a nerve-wracking moment: telemetry was lost just at the point of block DM ignition. In the event the DM lit automatically, burned for eight minutes and restarted twice again, each time bang on schedule. Its cargo, the 3,800 kg Direct TV-1R comsat, was at its geosynchronous destination 62 min later. Despite the novelty of the project, it attracted little media attention. What was probably more important for the promoters, though: the order book lengthened.

Despite this promising start to the venture, there was an unexpected setback on the second commercial launch in March 2000. Carrying a 2.7-tonne ICO F-1 comsat, the Zenit second stage shut down at 461 sec into the mission and the payload was lost,

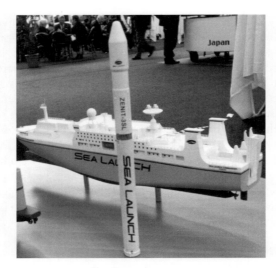

Sea Launch system

the Russians and Ukrainians blaming one another. A Russian–Ukrainian interstate investigation eventually settled the matter, attributing the cause to the software system.

This was yet another example of computer software programs being increasingly used to program complete launch operations, but where a minor error led to the complete loss of the mission. The classic original example was the maiden flight of the European Ariane 5 where the computer reset itself, thought the rocket was off-course, even though it was not and destroyed the vehicle. In this Sea Launch project, the computer program neglected to close a valve on the helium pressurization system on the second stage. As a result, pressure seeped away, the second stage lost thrust, closed down early and it failed to reach orbit. The real failure here was not so much the command that had not been included, but that checks of the software system had failed to spot the omission.

Despite this, Sea Launch did not lose any customers and was back in business within months. The Zenit 3SL soared into a dark blue summer sky on 28th July, delivering the PanAmSat 9 on station a mere 1 hr 45 min later. The transfer orbit was so accurate that the perigee was 1,900.5 km, compared with the 1,900-km target. Sea Launch had 17 customers on its books in autumn 2000 and the company was planning how to reduce the launch interval to 50 days in order to clear the waiting list.

March 2000 was the only mission loss, though there were problems with the 29th June 2004 launch, for the two block DM firings were 9 sec and 26 sec short, respectively, leaving Telstar with a perigee of 21,000, not the 36,000 km intended, with the shortfall to be made up by the Telstar 18's maneuvering engine.

Sea Launch went through a number of ups and downs. In the early years of the century, the commercial satellite market reached saturation and in 2002 there was only one launch. By this stage the project had still not re-paid its investment and there was some speculation that it might even close. The downturn did not last for long and the order book for Sea Launch began to lengthen again, to the point that Yuzhnoye began to give consideration to a more powerful version able to increase the weight of payload lifted from 6,000 kg to 7,500 kg. There were twenty Zenit 3SL launches over 2000–6, all but the first successful.

The idea of developing a land-based version of Sea Launch soon—predictably called Land Launch—was first mooted in 2001. The launch base would of course be Baikonour, where Zenit facilities already exist, saving on cost. Payloads would of course be smaller, around 3.6 tonnes to geostationery orbit, but could be offset, though, by the reduced costs of a land-based launch. Launch from Baikonour might therefore suit a smaller satellite, less heavy than the traditional weight carried by Sea Launch (3.6 tonnes). By 2006, Land Launch had attracted six contracts, with launches due to start 2007.

Zenit 3SL launches, 2000–6

12 Mar 2000	ICO F-1(fail)
28 Jul 2000	PanAmSat 9
21 Oct 2000	Thuraya
19 Mar 2001	XM-2 Roll
15 Jun 2002	Galaxy 3C
10 Jun 2003	Thuraya 2
8 Aug 2003	Echostar 9
30 Sep 2003	Galaxy 13
11 Jan 2004	Telstar 14 Estrela do Sul
4 May 2004	Direct TV-7S
29 Jun 2004	Apstar 5/Telstar 18
28 Feb 2005	XM-3 Rhythm
26 Apr 2005	Spaceway 1
23 Jun 2005	Telstar 8
8 Nov 2005	Inmarsat 4
15 Feb 2006	Echostar 10
12 Apr 2006	JCSAT 9
18 Jun 2006	Galaxy 16
22 Aug 2006	Koreasat 5
26 Oct 2006	XM-4 Blues

All from Odyssey platform

ROCKOT

Rockot was one of the the first post-Soviet new launchers, though, as may be expected, it had historic roots in the Soviet missile program. Rockot (code 15A35) was a small, 29-m-tall missile, based on Vladimir Chelomei's UR-100 missile. A total of 360 such missiles were deployed in silos around Russia and the Ukraine at the height of the Cold War; the Rockot was fired 144 times in long-range tests during the period. Now its manufacturers were searching for ways of disposing of the hardware peacefully and entering the commercial launcher market. Rockot was the first of the Cold War missiles to be adapted for civilian missions. Sub-orbital tests were made at Baikonour as early as 1990, before the first orbital mission in 1994.

The word Rockot means "roar of sound", not "rocket" (*raket* in Russian). Payload is in the order of 1.9 tonnes, with a commercial price of $14m for a launch. For commercial applications, a new third stage was added, called Briz KM. The KM was an adaptation of an old stage called Briz K, but with a relightable engine and a larger shroud. Briz KM weighed three tonnes, twice that when fueled. This was a versatile stage, which could fly for 7 hr and restart six times, dispensing different satellites into different orbits. Rockot's first stage uses a RD-233 motor, the second stage the RD-235. Fuels used are the traditional missile UDMH and nitrogen tetroxide. The diameter is 2.5 m and launch weight 107 tonnes.

Although there are several Rockot pads in northwest Baikonour, they were all underground silos and this presented a number of problems. The noise circulating in the silo during the first two or three seconds of liftoff is tremendous and there were fears that it would damage the satellite payload. At the same time as the Russian Space Agency was considering this problem, the Kazakh authorities got wind of the

Rockot and tower

Rockot launch out of Plesetsk

proposal and demanded a share of the profits of every Rockot mission out of Baikonour, with the result that the Russians moved all Rockot launches to Plesetsk.

This had an immediate cost, for it was now necessary to build an open, non-siloed pad at Plesetsk. To save costs, they converted an old Cosmos 3M launch pad and this involved the construction of a large vertical service structure, unusual in the Russian program. The Rockot is wheeled up to the structure in a vertical position, whereupon it is embraced by the service tower. Here, the payload is lifted by crane and placed on top of the bottom two stages. This was a procedure familiar to most space programs the world over (e.g., the United States, Japan, India, China), but was new to the rail-based Russians with horizontal assembly systems.

In 1995, Khrunichev formed a company with German Daimler-Benz Aerospace to market Rockot, offering it at €7m a launch. The company was then renamed Eurockot. Within five years, it had built up an order book of 12 launches for €200m. The company bought 45 old Rockots from the Russian strategic missile forces so as to build its inventory. In 2000, Eurockot was part-bought in turn by the German company Astrium GmbH, a shareholder of Arianespace (51%) and Khrunichev (49%).

Rockot was fired for the first time from Baikonour on 26th December 1994. It placed a Rosto radiosatellite into orbit, but the test was somewhat marred by the explosion of the third stage 3.5 hr after liftoff, scattering debris in a 2,000-km orbit at 64.6°. A second demonstration launch, this time entirely successful, was made in 1999 from the new pad at Plesetsk. In advance of flying its first commercial payload, Iridium satellites, mockup payloads called Simsat 1 and 2 were put into orbit in May 2000.

Rockot was launched eight times in 2000–6. Making a promising start, the first six launches were entirely successful. The seventh was a disaster, Cryosat, on 8th

October 2005. This was a high-profile launch carrying a European Space Agency satellite to survey the Arctic to measure global warning. The second stage should have stopped firing, dropped off and allowed the Briz KM third stage to ignite. Instead, the computer command to stop the second stage was not entered and the second stage burned to depletion. The third stage was commanded to fire only when the order to shut off the second stage was received, but, since this never happened, the third stage did not fire. The top-heavy rocket plunged back to Earth, crashing in the Lincoln Sea, just north of Greenland, ironically the very part of the planet that Cryosat was designed to study. In the confused and acrimonious aftermath, the Russian Space Agency stressed that the hardware did not fail and the fault was a human error in computer programming in NPO Khartron in the Ukraine, which made the software. The European Space Agency considered it to be such an important mission that it was decided to build another Cryosat, but the decision about a launcher was left open. The Russian side showed its determination to put things right, starting in traditional form with the dismissal of Khrunichev director Alexander Medvedev. New launch procedures were introduced, lines of management were straightened out to catch errors and the new Khrunichev chief, Viktor Nesterov, was required to report directly to the head of the Russian Space Agency, Anatoli Perminov.

Rockot was grounded until the following July, when it placed into a 685-km orbit the 800-kg South Korean Earth-observation satellite, Kompsat 2. The Russians were clearly on their best behavior, for the Koreans praised the level of service they received, encouraging the Rockot team to rebuild its order book [4]. Kompsat was an important mission for the Koreans. Using observation systems developed in Israel and Europe, it was Korea's first satellite able to send back photographs with a resolution of only 1 m.

Rockot launches, 2000–6

16 May 2000	Simsat 1,2
17 Mar 2002	GRACE 1,2
19 Jun 2002	Iridium 97, 98
30 Jun 2003	Monitor E model
	Mimosa
	Most
	Cubesat
	CUTE-1
	CanX-1
	AAU Cubesat
	DTUsat
	Quakesat
30 Oct 2003	SERVIS 1
26 Aug 2005	Monitor E
8 Oct 2005	Cryosat (fail)
28 Jul 2006	Kompsat 2

All from Plesetsk

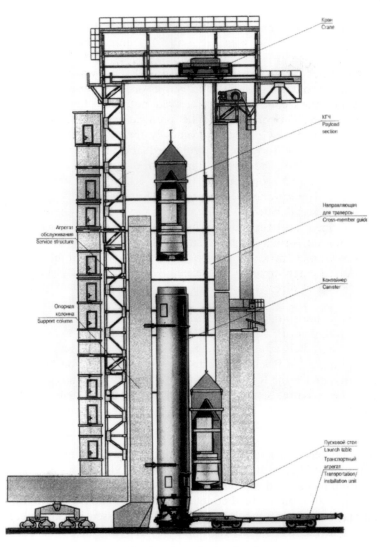

Кран
Crane

КГЧ
Payload
section

Направляющая
для траверсы
Cross-member guide

Контейнер
Canister

Агрегат
обслуживания
Service structure

Опорная
колонна
Support column

Пусковой стол
Launch table

Транспортный
агрегат
Transportation/
installation unit

Rockot launch site sketch

STRELA ROCKET

A new rocket was introduced in 2003 with the first orbital launch of the Strela (code 15A35). Like Rockot, Strela was a derivative of the UR-100 missile of the Chelomei design bureau and was a Cold War military missile known to the West as the SS-19, codenamed Stiletto. The UR-100 was 29.2 m long, 2.5 m wide, with two stages and joined the armaments in 1975, being installed in silos. Like Rockot, this was adapted with a view to commercial missions, the main change being that the multiple warhead

Rockot being raised

platform was modified with small engines to guide the payload into the correct orbit. Payload is up to 1,600 kg, with a commercial price of $10m.

In 2003 a dummy payload was put into orbit to test the system, achieving an orbit of 465 × 480 km, 67.08°, but no further details were given and Strela did not fly again. The payload may have been a mockup Kondor E remote-sensing satellite. Strela has two nitric acid and UDMH stages, each made by KBKhA in Voronezh.

Strela launchings, 2000–6
5 Dec 2003   Kondor model
From Baikonour

START

START is a converted solid-fuel military launcher derived from the SS-25 Topol missile and before that the SS-20 Pioner missile. The term "START" was an American acronym derived from the STrategic Arms Reduction Talks. Conceived in the

1960s by missile designer Alexander Nadirazhde (1914–1987), the Topol became an important part of the Soviet Union's nuclear strike force, with 288 missiles deployed in silos at nine locations, with the further ability to move the rocket around the countryside on mobile launchers. A successor, the SS-27 Topol M, now serves as the core of the small Russian silo-based nuclear strike force. There are two versions of START: START 1, the four-stage version, the only version flown in the new century and START 2, the five-stage version. START 2 made only one launch, an unsuccessful one of two small satellites (one Mexican, the other Israeli) in 1995 and was not used again.

Once the Cold War ended, the Moscow Kompleks Technical Center converted these missiles into modern, solid-fueled rockets which could place 700-kg-class satellites into orbit, with a commercial price around $10m. In a break with tradition for Russian satellite launchers, both were solid-fuel rockets. All other Soviet and Russian rockets are liquid-fueled, burning propellants and oxidizer from two tanks in an engine chamber. By contrast, solid-fuel rockets use a sludge-like gray substance that is poured into a single rocket tube: they are simpler and more powerful but less accurate and, once lighted, cannot be turned off, burning to depletion. Although solid fuels are an important part of the space programs of other countries (e.g., India, Japan, Europe's Ariane 5, the American shuttle), they played little part in rocket development in the Soviet Union. Solid-fuel rocket development came to a halt in the old USSR in the 1930s when the two designers most associated with solid-fuel technology, Georgi Langemaak and Ivan Kleimenov, were shot on Stalin's orders.

The first START 1 launch took place on 25th March 1993 from Plesetsk, placing a small test satellite in orbit and marking the first orbital solid rocket launch by Russia. START 1 was later the launcher used in the inaugural flight from Svobodny cosmodrome in 1997 of a small satellite called Zeya, named after the local river.

Over 2000–6, START was used three times, notably for two Israeli satellites and the Swedish Odin. The use of START by Israel may surprise, granted the less than warm history of the political relationships between the countries. Ironically, it arose from a 1995 incident in which a Mossad agent, Ruven Dinel, was caught red-handed receiving ten classified photographs in a metro station, taken from a Kometa mapping satellite, having suborned an official in its photographs archive. Russian military intelligence calculated that if the Israelis needed the pictures that badly, then there was scope for a commercial deal that would put the arrangement above board. Accordingly, Israel paid $15m for Russia to launch the Eros communications satellite, a project made easier by the fact that most of the participants on the Israeli side were emigrants from Russia, so there were no language barriers. Odin was a small 250-kg Swedish scientific satellite to study the atmosphere, including ozone depletion in the stratosphere.

START 1 launches, 2000–6
  5 Dec 2000   Eros A
20 Feb 2001   Odin
25 Apr 2006   Eros B
All from Svobodny

DNEPR

Dnepr was a converted Cold War rocket called the SS-19 Satan, able to deploy up to ten warheads of a total weight of three tonnes, so it was a formidable weapon of mass destruction. Built by the Yuzhnoye design bureau as the R-36M, it was deployed from 1974 and tested as a rocket 150 times. Production ceased in 1991, but most of the stock dated to the 1980s and 1970s. Like the Rockot, Dnepr was designed to be fired in rapid succession from silos. As a civilianized missile, Dnepr was able to put much larger, 4,200-kg payloads into Earth orbit; the price suggested by Yuzhnoye was $20m to $40m. Dnepr was operated by the Kosmotras company, which set a price of between $10,000/kg and $30,000/kg for a launch.

One hundred and fifty such missiles were left over at the end of the Cold War, with strategic arms limitation agreements and probably safety requiring their elimination by 2007. Under the Strategic Arms Reduction Talks agreements of 1991 and 1993, Russia committed to eliminating the Satan as an operational missile by 2001, but could use old versions to launch satellites. Both Kazakhstan and Ukraine gave their left-over Dneprs back to Russia for disposal.

The Dnepr had five silos at Baikonour, with pad 109 allocated to satellite missions. Dnepr was ready for a test mission by 1999. Its first payload was a small satellite made by the great British experts in small-satellite design, the University of Surrey. UOSAT 12 entered an orbit of 660 km, 64.5°, 97 min and carried a radio experiment, GPS receiver and Earth-observing cameras. Typically, the Dnepr would thenceforth carry a modest prime payload, with a number of small and very small satellites.

Dnepr found a niche as the launcher of small piggyback satellites. Dnepr was the first Russian launcher actually equipped to carry piggyback satellites. Until then, although piggybacks had been carried, they had been attached to the prime satellite, not the launcher and the prime satellite manager was responsible for ensuring deployment. With Dnepr, a small platform was fitted to the payload section to facilitate small-satellite deployment [5]. As an example, the launch of 1st July 2004 was centered around its core payload, the French Demeter satellite of 130 kg, designed to detect impending earthquakes, but it included five communications satellites of between 12 kg and 50 kg, a radio amateur satellite (12 kg) and a student research satellite for the University of Rome (12 kg).

For the future, Yuzhnoye plans to upgrade the Dnepr to the Dnepr 1, with a solid-fuel booster developed by NPO Iskra in Perm and a small upper stage called DU-802 developed within Yuzhnoye itself. The DU-802 has 4.5 kN of thrust and a specific impulse of 322.5 sec which could send payloads out of Earth orbit.

The Dnepr made five launches over 2000–6. The sixth, in July 2006, was a disaster, especially for Belarus, which had spent many years preparing Belka ("squirrel"), an Earth-observation satellite based on the Victoria platform developed by Energiya with two cameras, of 2.5 m and 10 m resolution. President Lukashenko traveled to Baikonour to watch the take-off. Also on board were a 92-kg student satellite called Baumanetz, named after the Bauman Institute where it was made, a University of Rome technology satellite with a camera, a reentry experiment and

Dnepr launch

global-positioning system and fifteen small satellites, ranging in weight from 35 kg to very small satellites of about 1 kg, made in numerous countries.

Heading toward the old railway station of Baikonour, the Dnepr flipped over about a minute into its flight, its remains making two large craters, one 25 km downrange, the other 150 km distant from the pad, near the settlement of Zhanakala. Smaller impacts were found later. The fuel of the second stage never ignited, so it fell back to Earth with the rocket, dumping 24 tonnes of heptil and 62 tonnes of nitric acid. Lukashenko returned to Belarus a disappointed man and there was sadness worldwide in the universities that had cargoes on board, ranging from Montana to Arizona, Korea to Japan. Some of the satellites were later found by helicopter teams near the old town of Baikonour: they were hardy and, although badly damaged, they had survived the crash back to Earth [6].

Kazakhstan made a big issue of the damage caused by the crash. Teams of ecologists were sent to fan out across the region and take soil samples, though high summer temperatures evaporated the deadly fuel and they had little to find. Zhana-kala was cordoned off and all the citizens were given health checks. Very dramatically, the regional prosecutor opened criminal proceedings. Eventually, Russia provided €1m compensation.

Russia appointed a commission of investigation immediately and its members duly descended on the two companies that made the first-stage engines, Yuzhnoye's Yuzhmash (Pivdenmash) factory in Dnepropetrovsk and the Kharton company nearby. They found that there had been a sharp increase in the fuel flow at 73 sec, which could be simulated if you took the insulation away from the fuel line. The commission of investigation reported in September, attributing the accident to an insulation failure on a line leading into the hydraulic engine drive on first-stage engine #4. The faster flow caused an interruption in thrust for 0.27 sec some 73 sec into the mission; even such a short interruption was sufficient for control to be lost. Instructions were issued for all lines to be checked for manufacturing defects and faulty insulation. Alexander Lukashenko had taken the precaution of insuring the satellite and issued orders for a new Belka 2.

Dnepr launches, 2000–6

27 Sep 2000	Megsat
	Unisat
	Saudisat 1A
	Saudisat 1B
	Tiungsat
20 Dec 2002	Unisat 2
	Saudisat 1C
	Rubin 3
	Latinsat 1
	Latinsat 2
	Trailblazer
29 Jun 2004	Demeter
	Unisat 3
	Saudisat 2
	Saudicomsat 1
	Saudicomsat 2
	Latinsat 5
	Latinsat D
	Amsat Echo
24 Aug 2005	OICETS
	INDEX
12 Jul 2006	Genesis 1
26 Jul 2006	Belka (fail)
	Baumanetz
	15 small satellites

All from Baikonour, except Genesis 1 from Dombarovska

VOLNA, SHTIL AND RELATIVES

Rockot, START and Dnepr were land-based missiles. The end of the Cold War left the Russian Federation with a surfeit not just of land-based missiles but also sea-based rockets, the equivalent of the American Polaris and Poseidon missiles. Several hundred were available in a number of different categories. The Makeev Design Bureau, based in Chelyabinsk, was the principal maker of these solid-fueled submarine-launched ballistic missiles. The main version was called the R-29 and was first tested from a submarine in the Black Sea in 1969, being declared operational in 1974, with a subsequent version, the R-29R, also called the RSM-50 but more popularly the Volna, the Russian word for "wave". A heavy version with a third stage, the R-29M, also called the RSN-54 but more popularly Shtil, was introduced in 1979 and each Delta 4 class submarine was equipped with sixteen such missiles. By the 1990s there were 112 left-over Shtils and similar numbers in other classes.

The idea of using the R-29 series for space missions dates to the 1970s, when the R-29R Volna sent a small recoverable cabin called Volane high into the atmosphere on scientific sub-orbital missions. Fourteen such missions were made [7]. This program was largely forgotten about until the 1990s when Makeev gave consideration to the use of submarine-launched missiles as a means of putting satellites into orbit. This had a number of advantages: it was a useful means of disposing of the old missiles, provided work for the grounded Russian navy and offered a ready-to-go launch pad in any suitable Delta submarine. Moreover, a submarine-launched satellite could be launched from any (watery) latitude on Earth, facilitating direct access to the intended orbit. They could be fired from the sea, over the sea, with no one underneath complaining about stages falling on top of them.

Makeev offered these missiles to Western companies and scientific bodies for orbital and sub-orbital flights. Although Volna and Shtil were the principal versions on offer—they were still in service on submarines—other variants were also proposed, such as Vysota, Skif, Surf, Priboi and Riksha. Surf was a different concept: the rocket was much bigger (104 tonnes) and could launch a much bigger payload (up to 2.4 tonnes). It was too large for a submarine tube and instead was to be towed into position in a container which floated vertically in the sea.

The only two used so far are Volna and Shtil, used on Delta 3 and Delta 4 class submarines, respectively. Volna had a payload capacity of 110 kg, Shtil up to 160 kg (more on future versions). The price for a Volna launch is about $1m, but more for a Shtil. Volna was 14 m long, 40 tonnes in weight and 1.9 m in diameter, Shtil slightly longer and thinner.

The first test launching from a submarine took place on 7th June 1995. Capt. Vladimir Bashenov fired the Volna rocket from 50 m below the Barents Sea near Yagelnaya Bay. This first test was a sub-orbital mission, with a small 700-kg capsule with a microgravity payload developed by the Bremen University Space Technology and Gravitation Center. The Germans were not allowed on board the submarine but helped to fit it at the naval base of Severomorsk. While arcing 1,270 km over Siberia, the Volna's capsule was able to provide almost 20 min of zero gravity for experiments into electrical fields and the behavior of fluids. It came down 30 min later, 5,600 km away in the far east. Data were transmitted downward to the Red Navy tracking stations in Severomorsk and Severodvinsk.

The Shtil's first orbital test came on 7th July 1998. The *Novomoskovsk*, under the command of Capt. Alexander Moiseyev and his 130 crew, dived off the coast of Murmansk and lay below the Barents Sea as the crew worked through the launching drills. The submarine then launched a Shtil missile toward orbit from below the Barents Sea. The rocket curved northward over Svalbard and Greenland, placing into 400×773-km, $78.9°$, 96.4-min orbit two tiny satellites from the Technical University of Berlin (TUB). Their cargoes were, accordingly, called Tubsats. Tubsat 1 was 8.5 kg and Tubsat N1 even smaller, 3 kg. The second successful launching was from the submarine *Ekaterinberg* on 26th May 2006. The Shtil had a 100% success rate on two missions.

The same cannot be said for Volna, which failed on all its four subsequent launchings. The first, in July 2001, was not an orbital attempt, but a sub-orbital demonstration of a solar sail in advance of a full mission, shooting 400 km high. It was intended to recover the sail satellite body with an inflatable device. In the event, the third stage failed to separate from the payload and the solar sail did not deploy. Further investigation suggested that the Volna underperformed: the computer sensed that not enough velocity had been reached and was programmed not to release the third stage until it had. So, it didn't.

On the second attempt to launch a solar sail, in October 2005, this time on an orbital mission, the circumstances of the loss remain a mystery. According to the Russians, the Volna stage stopped firing at 83 sec due to a turbopump malfunction. It seems that signals were picked up from the solar sail 6 min into the mission and some analysts believe that the failure took place at the final stage of entry to orbit instead.

On the other two Volna launches, which were for demonstration reentry vehicles, the demonstrators were never found (see Lavochkin in Chapter 7: Design bureaus). Although it is possible that the demonstrators themselves failed, most people have pointed the finger at the launcher.

Following the series of Volna failures, the head of the space agency, Anatoli Perminov, later recommended foreign organizations not to use the Volna again: it was simply too old and had been out of production for too long. A replacement is already in sight: the Bulava, first tested from the enormous Typhoon-class submarine *Dmitri Donskoi* from the White Sea on 27th September 2005. The Bulava is the submarine version of the Topol M missile and is slightly lighter, longer and thinner than the Volna (36 tonnes, compared with 40.3 tonnes; 16 m tall, compared with 14.8 m; 1.8 m diameter compared with 1.9 m). The Bulava was due to enter service on both the *Dmitri Donskoi* and the submarine *Alexander Nevski*, then completing construction in Sverodvinsk. While the Bulava has not yet been specifically kitted out for a satellite role, this would not prove difficult, granted that another version of the Topol has already been adapted as a START launcher.

A modernized version of the Shtil was prepared in 2006, called the Shtil 2, with a first sub-version called Shtil 2.1. The name Sineva has sometimes been attached to this series. The main feature was a much-improved payload weight, achieved by

modifying the payload shroud, removing the military mechanism for deploying warheads and installing a new nose cone. The first candidate mission was the 80-kg South
African satellite Pathfinder [8].

Submarine launchings, 2000–6

Date	Mission	Launcher	Submarine	
19 Jul 2001	Solar sail sub-orbit	Volna	*Borisoglebsk* (fail)
12 Jul 2002	IRDT 2	Volna	*Ryazan* (fail)
21 Jun 2005	Solar sail	Volna	*Borisoglebsk* (fail)
7 Oct 2005	IRDT 2R	Volna	*Borisoglebsk* (fail)
26 May 2006	KOMPASS 2	Shtil	*Ekaterinberg*

All from the Barents Sea

NEW ROCKET: ANGARA

Development of a new, large rocket began in 1995, the futuristic Angara. Approved
by the Russian government in 1994, first funding began to flow the following year. At
a time when ever-newer designs of prospective Russian rockets kept appearing at
international air and space shows, it was difficult to know whether Angara should be
treated any more seriously than the others—and it was not. However, the approval
announcement had two key phrases: the first was that this new launcher would be the
primary Russian rocket until 2030 and that pads would not be built at Baikonour.
Clearly, Angara was directed toward long-term Russian launcher independence, in
particular from Kazakhstan. The seriousness of the Angara project was apparent in
2000 when the first hardware appeared on the floor of the Khrunichev factory in
Moscow. Angara was clearly intended to be the mainstay of the Russian rocket fleet
for many years.

The Angara takes substantial advantage of existing technology. Angara builds
on the Zenit system and uses a development of the proven liquid-oxygen and
kerosene RD-171 Zenit engines for the first stage, though the engine is renamed
the RD-191. The second stage uses the Zenit's high-performance RD-120M engine,
the old RD-120, but with thrust raised from 82 tonnes to 93 tonnes. The upper stage
will be either the Briz M or the ultrapowerful KVD-1 motor developed in the course
of the Moon race in the 1960s. Angara 1.1 uses the Rockot nose fairing while the
Angara 1.2 uses the Soyuz ST fairing, which is in turn based on the Ariane 4.

One of the problems in reporting on the Angara rocket is the many design
evolutions through which it passed over the subsequent ten years (a record 18).
Currently, three main versions of the Angara are under construction, immemorably
called versions 1.1, 3 and 4. Agreement was reached between Russia and Kazakhstan
to develop a sub-version of the Angara 3 jointly, this version being called the
Baiterek. Angara will replace a number of rockets in the Russian launcher fleet:
the Angara 1.1 will take the place of the Cosmos 3M, Angara 3 the Zenit and Angara
4 the Proton. Although lighter than the Proton, the Angara has better performance

Angara 1

Angara 4

and can lift much heavier payloads. Working from the new pad (35) at Plesetsk, the two-stage version is to place between 26 and 28 tonnes into 63° orbits, and the three-stage version 4.5 tonnes into geostationary orbit. Angara will mark the most comprehensive program of rocket replacement in the space program, Soviet or Russian. It will reduce dependence on components made in the Ukraine and in the case of the Proton discontinue the use of toxic fuels.

Table 5.2. Versions of the Angara

Name	Wt	Capacity	Stages 1	2	3
Angara 1.1	145	1.7 LEO	RD-191M	Briz M	—
Angara 3	464	15.2 LEO, 4.5 GEO	RD-191M	RD-0124	Briz M
Baiterek	480	14 LEO, 4 GEO	RD-191M	RD-0124	KVD-1
Angara 4	752	28 LEO, 7.3 GEO	RD-191M	RD-0124	KVD-1

Table 5.2 summarizes the capacities of the Angara versions currently planned (the names and figures have undergone numerous changes and they may continue to do so). Although they present as one launcher family, the capacity of the four versions varies enormously. Angara 1.1 is a lightweight launcher while the Angara 4 is a good 25% more powerful than the Proton.

The Angara 1.1 is expected to be launched first. Like the other members of the series, it is expected to be promoted commercially and the price for the Angara 1.1 is estimated as $20m. In summer 2005, the Defence Ministry signed a contract with the

Angara launch site sketch

RD-191 engine

Khrunichev company for six Angara launchers to be built and delivered by 2010. Three were for three medium versions (Angara 3) and three for heavy versions (Angara 4).

There is also a proposal for a reusable, winged version of the 1.1 called the 1.2. After staging, the first stage sprouts wings and turns into a glider, coming back to Mirny airfield where it is safed, any remaining propellants removed and preparations begin for the next mission. Also, borrowing systems from elsewhere, Angara 1.2 uses the undercarriage of the Sukhoi 17 jet-fighter and a maneuvering jet engine taken from the Yakovlev 130 jet. It is not known how viable this system is: it will probably be developed last of the series.

RUSSIAN ROCKET ENGINES

Chief designer Sergei Korolev once remarked that at the heart of a good space program is a good rocket engine. Engines are the single most important element in a rocket's design and are crucial for high performance and reliability. Was Russia able to maintain and develop the achievements of the Soviet Union in rocket engine design?

Rocket engines were very much the starting point of the entire Soviet space effort, dating back to the foundation of the Gas Dynamics Laboratory in the 1920s. Engine design was made a priority from the very start and the engine design bureaus attracted the most talented of the Soviet Union's engineers. Leadership of the space program often came from the engine design bureaus, most evident when the greatest engine designer of them all, Valentin Glushko, became the chief designer of the entire program from 1974, a post which he held till his death in 1989.

Russia's rocket engines have been designed in four main bureaus. The most important one is the Gas Dynamics Laboratory (GDL), now NPO Energomash; the others are the Kosberg bureau (KBKhA), the Isayev bureau (KhimMash) and the Kuznetsov bureau. Their work and role in the Russian space program are reviewed (Table 5.3 lists the main rocket engine types).

Table 5.3. Engines in the Russian rocket fleet

	1	2	3	4
Proton	RD-253	RD-0210	RD-0212	11D58M
Proton M	RD-275	RD-0210	RD-0212	Briz M
Soyuz (U version)	RD-107	RD-108	RD-0110	
Molniya M	RD-107	RD-108	RD-0107	S1 5400
Shtil	RD-0243	NI	NI	NI
Rockot	RD-0233	RD-0235	Briz KM	
Strela	RD-0233	RD-0235		
Dnepr	RD-251	RD-0255		
Cosmos 3M	RD-216M	11D49		
Tsyklon 2/M	RD-261	RD-262		
Tsyklon 3	RD-251	RD-252	RD-861	
Zenit 2	RD-171	RD-120		
Zenit 3SL	RD-171	RD-120M	11D58M	
Angara*	RD-191M	RD-120	Briz M	

Names where known.
Data not available for START-1, Volna, Strela.
NI = Not identified.
Depends on stage.

GDL/ENERGOMASH: THE MOST POWERFUL ROCKETS IN THE WORLD

The most famous rocket engine design bureau was the Gas Dynamics Laboratory, founded in Leningrad soon after the revolution, whose history has been intimately associated with Valentin Glushko. As a design bureau, it went through many evolutions, becoming merged with the main design bureau, Korolev's OKB-1, when Glushko became chief designer (1974), but going its own way again as NPO Energomash (1990), its current name, or, to be more precise, *Energomash imemi Valentin Glushko* ("Energomash, dedicated to the memory of Valentin Glushko"). Privatized in 1998, it is probably still the biggest engine design bureau, with a floor area of $282,000\,m^2$ and 6,500 people still work in its Khimki headquarters and subsidiaries [9]. One could say that it is the world's foremost rocket engine design bureau, having produced a total of 52 operational engines during the 20th century. Its engines are considered the best in the world—not just by the Russians, as one might expect—but by the Americans who are now using them to power their new generation of Atlas rockets.

His engines may be divided into six groups: the RD-100 series, which use liquid-oxygen and non-storable propellants; the RD-200 series, which use nitrogen and storable propellants; the RD-300 series of flourine engines; the RD-400 series of nuclear engines; and the RD-700 series of tri-propelllant engines. RD stands for rocket motor (*raketny dvigatel*).

In the 1950s, GDL designed the RD-107 and RD-108 liquid-kerosene-fueled engines for the R-7 rocket, used initially to launch Sputnik and since then developed in the Vostok, Molniya and Soyuz versions. The RD-108 was used as the core stage on the R-7 (block A), the RD-107 on the four side stages (blocks B, V, G, D).

Valentin Glushko

RD-108

Each RD-107 weighed 1,155 kg dry and had a thrust of 1,000 kN. The RD-107 (and its relative, the RD-108) must be the most used rocket engine in history, for with 1,716 launches since 1957, each firing 20 motors at liftoff, this makes a total of 34,000 engines! They have been made and still are made by the Motorostroitel plant in Samara.

In the early 1960s, GDL developed engines with storable propellants for the missiles built by the Soviet Union, and some of these were adapted for civilian use, principally in the Cosmos rocket (RD-214, RD-216) and the Tsyklon rocket (RD-218, 219, 251, 252, 261, 262). Nitrogen-based storable fuels were much more suited to missiles than the R-7, for they could be kept at room temperatures for long periods, but they had the disadvantage of being toxic and required careful handling.

Energomash headquarters

Glushko believed that he could achieve much more power with nitrogen-based fuels, but chief designer Sergei Korolev loathed them and wouldn't use them. Rocket engines were dangerous enough anyway, but he called these fuels "the devil's own venom".

In the mid-1960s, Glushko developed closed-cycle engines, using the gases generated by the engine's thrust to power the engine itself. The main breakthrough was with the RD-253, the storable fuel engine used to power the Proton rocket and which first flew in 1965. The RD-253 was one of the most powerful engines of its day, delivering pressures of hundreds of atmospheres. Each engine weighed a modest 1,280 kg. The turbines went round at a fantastic 13,800 revolutions a minute or 18.74 MW. Temperatures reached 3,127°C in the engine chambers and their walls were plated with zirconium. Their specific impulse was 2,795 m/sec at sea level and 3,100 m/sec in vacuum. The engines are made in the Sverdlov plant in Perm in the Urals and about 1,800 have now been used. For the Proton M, introduced in 2001, the engines were improved to achieve an extra 7% thrust, 162 tonnes against 150, 163 atmospheres against 150. The improved version is called the RD-275. Later, there will be a further improvement, the RD-276, with 5.3% more thrust, 170.4 tonnes; they ran for 735 sec of tests in 2005.

The flagship engine of Energomash is the RD-170, developed in the 1970s for the Energiya system and its Zenit strap-on rockets (where the version is called the RD-171). With Energiya, Glushko was obliged to return to kerosene. One of the Proton's early spectacular failures was in April 1969, when one blew up soon after take-off on a Mars mission. A large number of generals was in attendance and there was nowhere to run from the falling nitric acid, nor any way of disposing of it until the next rain washed it away into the ground. Badly shaken, they ensured that the subsequent set of rocket design studies, called *Poisk* and carried out over 1973–74, decided that future rockets were to use conventional fuels.

Returning to kerosene, the RD-170 weighs 8,755 kg, has four chambers and has a thrust of 7,905 kN or 740 tonnes. The engine can be throttled in a range from 40% to 105%. Its weight is 13 tonnes, height 4.3 m and diameter 4.1 m. An indication of the progress of rocketry may be gauged from the fact that chamber pressure rose from 16

Boris Katorgin

atmospheres on the RD-100 to 250 atmospheres in the RD-170, the highest thrust of any Soviet engine.

The RD-170 went through a difficult design history, the first sixteen tests in 1980 failing due to high vibration and burn-throughs [10]. An RD-171 destroyed its test stand in Zagorsk the following year, when it exploded 6 sec into a test. The problems were so severe that the abandonment of the RD-170 program was contemplated, with three alternative engines considered. In the end Energomash chief designer Valentin Glushko and NPO Yuzhnoye chief designer Vladimir Utkin decided to persevere, achieving successful test firings with the RD-170 in 1983 and the RD-171 the following year. The RD-170 performed perfectly on the two flights of Energiya, as did the RD-0120 hydrogen-powered upper stage.

A version of the RD-170, but with only a single chamber, will be used to power the new Angara rocket; it is called the RD-191. It is a 2.2-tonne engine able to generate over 196 tonnes of thrust. The RD-191 completed assembly in April 2001 and made its first test on 27th July 2001, the first step in a two-year qualification program. Four tests were made by the end of the year and in 2002 firings of up to 150 sec were made. By 2006, 13 stand firings had totaled 1,028 sec.

The RD-100 and 200 series have been the main line of development in Energomash. Some rockets were built in the 300 series: the best example was the RD-301, an experimental engine developed in 1969–76 using fluorine and ammonia, generating a thrust of 98 kN for 750 sec, but it was not used operationally. Energomash also made a limited exploration of nuclear engines (the RD-400 series) over 1958–63 and again in the 1980s, but not since. Current director is Boris Katorgin.

RD-180 POWERS THE ATLAS

The power of Russian rocket engines had always been recognized abroad. Once the Soviet period ended, American aerospace companies were quick to establish commercial ventures with engine designers and manufacturers. In 1992, NPO Energo-

RD-180 engine

mash signed a preliminary agreement with America's prime engine manufacturer, Pratt & Whitney, to market its engines in the USA. In early 1996, a version of the RD-170 engine, called the RD-180, won the competition to power the new American Atlas rocket, designed to be the main conventional booster rocket for the United States from 2000. The first production models were then shipped to the United States and over a hundred were ordered.

The RD-180 export turned into a great success story. Lockheed Martin, engaged in competition with Boeing for the American launcher market, bought 101 RD-180 engines from Energomash for $1bn. The program for fitting the RD-180 to the new Atlas rocket was overseen by a joint company called Amross. Despite the cooperative nature of the venture, the Americans applied national security restrictions rigorously. The 25 Russian engineers overseeing the first launch of the new Atlas were not allowed in the Cape Canaveral control center but made to follow the launch in an adjacent hangar under the supervision of the U.S. Defence Threat Reduction Agency.

The Atlas III was designed to replace the Atlas II, which had been the workhorse of medium-size American satellites for many years. Ironically, it was based on the original Atlas which had been targeted on the Soviet Union since the late 1950s. Now the Atlas was to be reequipped with the engines of its adversary, both for the Atlas III and a later model, the Atlas V.

RD-180 on Atlas V

The RD-180 was so powerful that its one motor replaced the two engines of the Atlas II—and provided 30% more thrust. The RD-180 was a two-chamber version of the RD-170, giving some idea of the capacity of the 170 itself. Not only that, but it was the first American launcher, apart from the shuttle, that could throttle its engines, giving it another performance advantage. The throttle range was from 37% to 100%.

The RD-180 had 15,000 fewer parts than previous Atlas engines, making a failure much less likely. Its use of oxygen enrichment in the early stages of the burn enhanced performance and kept running temperatures down.

After a string of first-launch failures, such as the European Ariane 5, Lockheed Martin engineers were naturally apprehensive as the first flight of the RD-180-powered Atlas III neared. Their nerves were not steadied by four aborted countdowns, called off due to weather, radar failures and, most irritatingly of all, pleasure boats straying up the Canaveral coast to the launch site.

By the time the first Atlas III had counted down to zero on 24th May 2000, the engine had already reached twice the engine pressure of any previous American launch vehicle, 3,700 psi. At liftoff, the RD-180 was on 74% thrust, generating 288,716 kg thrust, kept deliberately low so as not to damage the launch complex. Only 5 sec later, the RD-180 had accelerated to 92% thrust, burning an oxygen-rich mixture at the rate of a tonne of oxygen every second.

Next came the crucial stage of maximum dynamic pressure, or max Q as the engineers call it. At this stage, the vehicle goes supersonic as it pushes through the densest layers of the atmosphere. Pressures on the launcher are so intense that there is a real danger of the vehicle breaking up—indeed, it was at this very point that the *Challenger* exploded in 1986. Now, at 33 sec into its mission, the twin-nozzled RD-180 throttled back to 64% of thrust until 63 sec as the vehicle went through this difficult phase. A minute after take-off, through the low atmosphere, the RD-180 was up to 87% and now rapidly accelerating skywards. In three minutes it had reached the same speed and altitude as the Atlas II had in five minutes. It was soon performing well over its 362,880 kg rated thrust. The Centaur upper stage then took over to bring the payload to geosynchronous orbit. The first-time success of the RD-180 left Lockheed and Energomash ecstatic.

In August 2002, it was followed by the more powerful Atlas V, which put the Hotbird 6 broadcast satellite into orbit. This was ultimately more important, for the Atlas V was to become one of two heavy-lift American lifters for the new century, the Atlas III being retired in 2005. The Atlas V's Russian engines were used to send the Mars Reconnaissance Orbiter to the red planet in August 2005.

KOSBERG BUREAU/KBKHA IN VORONEZH

Known as KBKhA, KB KhimAutomatiki, in English the Chemical Automatics Design Bureau (CADB) worked initially on aircraft engines and then, from around 1956, rocket engines. The engines designed by Semion Kosberg (1903–65) are used for the upper stages of Russian rockets. Kosberg's design bureau was set up in Moscow in 1941, evacuated to Berdsk and then settled in the city of Voronezh after the war. The current anodyne name of "chemical automatics" is not one loved by its current workers (it was probably another Soviet terminological deception, designed to distract Western analysts from its true purpose) and it is more commonly referred to as the Kosberg bureau.

The Kosberg KB developed the engines for the upper stages of the R-7, namely the RD-0105 engine for the Luna probes, the RD-0109 engine for the Vostok upper stage, the RD-110 engine for the block I (Soyuz) and the RD-0107 engine (block L, Molniya), all of which used liquid oxygen and kerosene; and the upper stages of the Proton, namely the block B (RD-0210) and block V (RD-0212), using nitrogen tetroxide and UDMH, which gave so much trouble in the late 1960s. KBKhA had substantial experience in the development of nuclear engines in the 1980s and remains in the forefront of rocket engine design. It has also built a substantial number of engines designed and developed elsewhere, such as the Briz.

KBKhA was responsible for the new upper-stage engine of the Soyuz 2.1.b, introduced in 2006. Chief designer Vladimir Rachuk pointed out that the RD-0124 was the first entirely new upper-stage engine design introduced in the post-Soviet period (the others were adaptations rather than fresh designs).

ISAYEV BUREAU/KHIMMASH

The Isayev bureau was set up in 1943. The bureau started life as plant 293 in Podlipki in 1943, directed by Alexei M. Isayev and was renamed OKB-2 in the 1950s, being given its current name, KM KhimMash in 1966. His engines were used in the rocket program as small-propulsion system, maneuvering and orientation engines. Besides spacecraft, its work has concentrated on long-range naval, cruise and surface ballistic missiles and nuclear rockets and by the early 1990s had built over 100 rocket engines, mainly small ones for upper stages, mid-course corrections and attitude control. Examples were the KDU-414 (used on the early planetary probes), the KTDU-1 (employed on Vostok), the KTDU-5A (used for the Luna soft landings), the KTDU-53 (for Zond), the KRD-61 (for the Luna 16 ascent stage) and the KTDU-35 (used for Soyuz). The Isayev bureau made the systems of the Soyuz T series and the thrusters used on Salyut and Mir. The bureau made the KVD-1 rocket engine for the Moon program, whose derivative, the KVD-1M, powers the Indian GSLV (see Chapter 7).

Two other rocket engine bureaus should be mentioned. First, electric engines are made by the Fakel Design Bureau. The Soviet Union devoted some attention to the development of electric engines, flying two Arsenal-built Plazma spacecraft in the 1980s for lengthy tests of electric engine systems [11]. This work continued and Fakel's two main engines, the SPT 70 and SPT 100, are used by 24-hr satellites for long-duration station-keeping (up to ten years)—for example, those in the Ekspress series. Work is in progress on new electric engines, such as the SPT-140 and SPT-200. Second, an emerging contributor to rocket engine development is the Keldysh Center. This establishment dates to the 1930s and was directed from 1946 to 1961 by one of Russia's greatest scientists, Mstislav Keldysh, after whom it is now named. Located near Khimki, its emphasis has been on research and development; it is a growing participant in cooperative programs with Europe.

Plazma spacecraft (Cosmos 1818, 1867)

AND FROM HISTORY, KUZNETSOV'S NK-33

A final rocket engine, a ghost from the past, still haunts the Russian rocket engine program, the NK-33. This is a very old engine, built for the Moon program in the 1960s by the Kuznetsov Aviation Design Bureau then headed by Nikolai Kuznetsov (1910–95). When the manned Moon program was canceled by chief designer Valentin Glushko in 1974, he ordered that all the relics from that period, including the NK-33 engine, be destroyed. The Kuznetsov engineers in Samara could not bring themselves to carry out what they regarded as technological vandalism, so they acknowledged the order and hid the engines away in a hangar—hundreds of them (at least 450). They even placed skull-and-crossbones "keep away!" anti-radiation signs over them to deter curious prying eyes.

The engines were rediscovered in the 1990s, almost by chance, by a visiting group of American engineers. They could not believe what they saw—hundreds of unused, high-performance rocket engines, all in mint condition, gathering dust. The American Aeroject company bought some and sent them off to its Sacramento, California plant for testing and evaluation. They worried that there would be problems in relighting motors that had been in storage since 1974. They ran two tests—of 40 sec and 200 sec—and there were not. Aerojet's evaluation of the engine found that it could deliver over 10% more performance than any other American engine and enthused over its simplicity, lightness and low production costs.

Ultimately, the RD-180 won the battle to be the Russian engine to power the American rocket fleet. The NK-33, though, has never gone away and repeatedly features in proposals for new domestic and foreign launchers, having the great advantage that it has been built, tested and performs. Now, under the Rus-M program to upgrade the R-7 launcher, the engine is proposed to power the new Soyuz 3 rocket, so maybe this venerable 1960s' engine will fire after all, sixty years later.

FUTURE LAUNCH VEHICLE AND ENGINE PROGRAMS: URAL, BARZUGIN

The main design bureaus continue to explore the possibilities of new rocket engines. Energomash, KBKhA (Khim Automatiki) of Voronezh, KB KhimMash of Korolev have continued their design work on improvements to be adapted to existing rockets or prospective ones such as Onega [12].

The main development program is called Kholod, intended to explore the potential of scramjets that could be used in the atmosphere on future shuttle-type launchers. This involved the testing of a scramjet and involved the Gromov Flight Center, KBKhA Voronezh, TsAGI (Central Institute for Aero Hydrodynamics) and the Raduga Design Bureau, along with some support from NASA and the French space agency CNES. Five launches were made over 1992–98, reaching Mach 6.5, an altitude of 27 km and an operating time of 77 sec. The next stage is to test an 8-m-long liquid-hydrogen scramjet engine called Igla ("needle" in English) on a Strela rocket sub-orbital Mach 14 test downrange to Kamchatka; this is scheduled for 2009. On 18th February 2004, a military test along these lines was carried out on a Strela sub-orbital flight from Baikonour. This could have been the first free flight of an experimental scramjet, making Russia a world leader in this area of development [13].

The main area of future development is likely to be in the context of the European Space Agency's Future Launchers Technology Program (FLTP) and Future Launcher Preparatory Program (FLPP), begun in the early 2000s to look at launcher possibilities to come after the Ariane 5, called new-generation launchers. A modest budget was proposed, €40m.

France moved ahead, setting aside €200m for five years for development work, effectively its contribution to FLPP. This took the form of an accord dated 15th March 2005 between the French space agency, CNES and the Russian space agency, Roscosmos. This program, called Ural (or, in French, Oural) was designed both to make studies and also ground-test new systems and materials [14]. The reason for the high level of French interest was that most of the Ariane rocket systems are made in France by French companies and these faced a serious problem of holding on to their engineers and developing new lines of work in the period after Ariane. Put another way, France stands to lose most, if future European launchers are not developed. Ural had five fields of work: launch vehicle concept studies, the examination of methane propulsion, advanced cryogenic tanks, first-stage demonstrators and reentry vehicle demonstrators. The main companies involved on the French side were Astrium, Cryospace and Snecma while on the Russian side the participants were the Keldysh Center and TsNIIMash.

Key elements of Ural and related cooperation in the FLPP were:

- The Volga program: Snecma working with the Keldysh Center, Energomash and KhimAutomatiki on an oxygen–methane engine with 200 tonnes thrust and relightable 50 times. This also drew in the large European aerospace company EADS in Germany, Techspace Aero (Belgium) and Volvo (Sweden).
- EADS with KhimAutomatiki, to test a reusable next-generation engine.
- EADS with Khrunichev and the Keldysh Center on ion engines.
- Pré-X, a lifting body which could be tested by the European Vega or the Russian Dnepr launcher, involving CNES, the French defence development agency ONERA, EADS, Snecma, Dassault, Russia and in Germany MAN Technologie.

- Flex, a first-stage reusable demonstrator involving France, Germany and Russia (Khrunichev, Energiya and NPO Molniya).
- Structure X, ground tests of the materials needed for hydrogen engines, with a program of work being undertaken by France (Air Liquide, Cryospace), Germany, Russia and Switzerland.

Good progress was made on Ural in the first year [15]. Methane-powered engines had long been a theme of Russian rocket engine development and the Ural program adapted the hydrogen-powered KVD-1 engine, originally intended for the Moon landing in the 1970s, as a methane-fueled engine. Called the KVD-1.2, it made a 17-sec test run in December 2005 [16].

The Barzugin study was a concept study of a future development of the Ariane 5, replacing its two side solid-fuel rocket boosters with two liquid-propelled flyback rockets [17]. Using existing technologies, it offered the prospect of an extension of the Ariane 5 into the 2020s with significant cost savings through the use of reusable boosters. The technique of recovering side boosters is already used by the American space shuttle, for its solid rocket boosters parachute into the sea for recovery by two NASA ships operating out of Port Canaveral. Here, though, the Barzugin boosters would glide back to the runway at Kourou, saving them a drenching in corrosive, salty seawater. Detailed outline designs were completed by autumn 2006.

According to French engineers, the Ural and Barzugin programs are now at the heart of French–European–Russian cooperation [18]. They combine Europe's double need for an Ariane replacement and to develop its engine technology with Russia's experience in rocket engines and need for foreign partnerships bringing funding. They appear to be more, though, than just a system for technical cooperation, but an attempt to build up a new, durable and strong driver of space development.

Finally, a word on the testing of rocket engines. The main center for the test-firing of rocket engines is Sergeev Posad. The NIIKimash rocket test center was set up in Novostroyka in Sergeev Posad in the late 1940s (then known as Zagorsk), although it operated under the cover of a machine-building center for the chemical industry. Surrounded by 17 km of barbed wire fences, the test center was built on ravines overlooking the river Kunya. All rocket engines for the Soviet and Russian space programs have been tested there. Sergeev Posad continues in operation.

RELIABILITY

There were 13 Russian launch failures over 2000–6. The following is a list of Russian failures. These thirteen failures comprised four Volna, two Proton and the rest one each for the Zenit 3SL, Cosmos 3M, Tsyklon 3, Rockot, Soyuz U, Molniya M and Dnepr. During the period there were 192 launches, so these failures must be set against a reliability record of 93.6% of all launches.

They failed ... launch failures 2000–6

12 Mar 2000	ICO F-1	Zenit 3SL	*Odyssey* platform
21 Nov 2000	Quickbird 2	Cosmos 3M	Plesetsk
27 Dec 2000	Three Gonetz, Strela	Tsyklon 3	Plesetsk
19 Jul 2001	Solar sail sub-orbit	Volna	Barents Sea *Borisoglebsk*
12 Jul 2002	IRDT 2	Volna	Barents Sea *Ryazan*
15 Oct 2002	Foton M-1	Soyuz U	Plesetsk
25 Nov 2002	Astra 1K	Proton K	Baikonour (reached orbit)
21 Jun 2005	Molniya 3K	Molniya M	Plesetsk
21 Jun 2005	Solar sail	Volna	Barents Sea *Borisoglebsk*
7 Oct 2005	IRDT 2R	Volna	Barents Sea *Borisoglebsk*
8 Oct 2005	Cryosat	Rockot	Plesetsk
1 Mar 2006	Arabsat	Proton M	Baikonour (reached orbit)
26 Jul 2006	Belka	Dnepr	Baikonour

There were no failures in 2003 or 2004. 2005 was the worst year, with four failures, for some reason taking place in pairs. To make things worse for Russia, two were foreign missions. This level of failure put Russia in an unfavorable position compared with the two countries with the most reliable launch rates, the United States and China. It is difficult to try to compute in comparable failure rates for Europe, India, Israel and Japan, because of their low rate of launches.

One approach is to try to isolate those rockets that have presented the main problems of reliability (Table 5.4).

Table 5.4. Reliability of Russian rockets by rocket

	Launches	Failures	% Reliability
Soyuz U	878	20	97.7
Soyuz Fregat	4	0	100
Soyuz FG	13	0	100
Molniya M	293	19	93.5
Soyuz 2	4	0	100
Proton	323	37	88.7
Proton M	14	1	93.9
Zenit 2	36	8	77.7
Zenit 3SL	22	1	95.5
Cosmos 3M	438	23	94.7
Tsyklon 2/M	105	1	99
Tsyklon 3	120	8	93.3
Rockot	10	1	90
Dnepr	7	1	85.6
START (all versions)	7	1	85.6
Volna	5	4	20
Shtil	2	0	100
Strela	1	1	100

Based to end 2005: ESD, 2006.

All these failures had different causes and it was difficult to establish a common pattern. Two explanations do suggest themselves. First, several of the rockets used were quite old and had spent long periods in storage, in the course of which they may have deteriorated. This may contribute to explaining the problems with Cosmos 3M, Tsyklon, Volna and Dnepr. Second, it may well be the case that, as a result of funding shortages, some quality control may have suffered. Although the United States have achieved an almost faultless reliability record, this may be attributed to infinitely more generous budgets (see Chapter 7).

CONCLUSIONS: ROCKETS AND ROCKET ENGINES

The period from 2000 to 2006 marked a number of important developments in the Russian rocket fleet and in the development of rocket engines. Principal among these were:

- the continued use and the upgrading of R-7 variants, with the introduction of the Soyuz FG and the Soyuz 2;
- the commencement of the Proton M;
- the introduction of new upper stages, principally the Fregat and the Briz M;
- the retirement of the Tsyklon 3, the approaching retirement of the Tsyklon 2 and the gradual retirement of the Cosmos 3M;
- launches by a number of Cold War rockets, notably the Dnepr, Rockot, START, Shtil and Strela;
- the success of the RD-180 as an engine for American rockets;
- the prospect of the introduction of the Angara;
- the commercial success of the Proton and Zenit 3SL.

The Angara holds out the prospect of rolling over much of the old fleet and its replacement. The successful development of the RD-190 engine gave cause for hope that this could take place in the next few years, while the cooperation program with France, Ural, held out the best prospects for using to the full Russia's vast experience in engine development.

REFERENCES

[1] For the permutations of the Aurora, Onega and Yamal series, see Hendrickx, Bart: In the footsteps of Soyuz—Russia's Kliper spacecraft, from Brian Harvey (ed.): *2007 Space exploration annual*, Springer/Praxis, 2006.

[2] Degtyarev, Alexander and Ventskovsky, Oleg: *Yuzhnoye prospective design systems*. Paper presented to the International Astronautical Federation, Valencia, Spain, October 2006.

[3] Hendrickx, Bart: The origin and evolution of the Energiya rocket family. *Journal of the British Interplanetary Society*, vol. 55, #7–8, July–August 2002.

[4] Taverna, Michael: KARI on—Kompsat 2 launch provides boost to Korean space program as Rockot reenters service. *Aviation Week and Space Technogy*, 7th August 2006.

[5] Webb, Gerry: *The use of former Soviet Union launchers to launch small satellites.* Paper presented to the British Interplanetary Society, 5th June 2005; see also Pirard, Théo and Lardier, Christian: Kosmotras: une affaire qui marche. *Air & Cosmos*, #1990, 1 juillet 2005.

[6] Oberg, Jim: *Space officials seem to revert to Soviet ways after crash in Kazakhstan.* Posting on MSNBC, 14th August 2006.

[7] Pillet, Nicolas: *Les lanceurs dérivés du missile R-29: Volna et Shtil. http://membres.lycos.fr*

[8] Webb, Gerry; Degtyar, Vladimir; Sleta, Alexander and Sokolov, Oleg: *Shtil 2.1—a small satellite launcher with improved utility and evolutionary potential.* Paper presented to the International Astronautical Federation, Valencia, Spain, October 2006.

[9] Siddiqi, Asif: Rocket engines from the Glushko design Bureau, 1946–2000. *Journal of the British Interplanetary Society*, vol. 54, #9–10, September–October 2001.

[10] Hendrickx, Bart: The origin and evolution of the Energiya rocket family. *Journal of the British Interplanetary Society*, vol. 55, #7–8, July–August 2002.

[11] Grahn, Sven: The US A program and radio observations thereof. Posting on *www.svengrahn.pp.se*

[12] Lardier, Christian: Les nouveaux moteurs fusées russes. *Air & Cosmos*, #1807, 31 août 2001; Les nouveaux moteurs de CADB. *Air & Cosmos*, #1845, 21 mai 2002; Nouveaux moteurs russes et ukrainiens, *Air & Cosmos*, #1823, 21 décembre 2001.

[13] Platt, Kelvin: Russia's hypersonic engine development—a world leader. *Journal of the British Interplanetary Society*, Space Chronicle series, vol. 58, supplement 2, 2005.

[14] Lardier, Christian: Coopération spatiale euro-russe. *Air & Cosmos*, #1902, 5 septembre 2003; Accord sur le program Oural. *Air & Cosmos*, #1975, 18 mars 2005.

[15] Astorg, Jean-Marc; Louaas, Eric and Yakuchin, Nikolai: *Ural—cooperation between Europe and Russia to prepare future launchers.* Paper presented at the International Astronautical Federation, Valencia, Spain, October 2006.

[16] Bonhomme, C.; Theron, M.; Louaas, E; Beaurain, A. and Seleznev, E.P.: *French–Russian activities in the LOX/LCH4 area.* Paper presented at the International Astronautical Federation, Valencia, Spain, October 2006.

[17] Sumin, Yuri; Kostromin, Sergei; Panichkin, Nikolai; Prel, Yves; Osin, Mikhail; Iranzo-Greus, David and Prampolini, Marco: *Development of the Barzugin concept for Ariane 5 evolution.* Paper presented to the International Astronautical Federation, October 2006.

[18] Godget, Olivier; Arnoud, Emile; Prampolini, Marco; Prel, Yves; Talbot, Christophe; Kolozezny, Anton and Sumin, Yuri: *Next generation launcher studies—preparing long term access to space.* Paper presented at the International Astronautical Federation, Valencia, Spain, October 2006.

6

Launch sites

Chapter 6 looks at the cosmodromes used to service the Russian space program and also at its recovery and ground facilities. Russia now has nine cosmodromes—active, disused or planned—starting with:

- Kapustin Yar, on the banks of the Volga, where German A-4s were first tested, also known as the "Volgograd station". It is rarely used now.
- Plesetsk, in the north near the town of Mirny, originally a missile base and then used mainly for military launchings.
- Baikonour, in Kazakhstan, the best known, from which manned, lunar, inter-planetary and geosynchronous missions are launched.

In recent years, new cosmodromes have been developed:

- Svobodny Blagoveschensk, an old missile base in the far east, for small satellites.
- Yasny/Dombarovska, in the southern Urals, another old missile base.

In addition, two maritime sites were used:

- The Barents Sea, for submarine-launched small satellites.
- The *Odyssey* platform in the Pacific Ocean, for the Sea Launch project.

and two launch pads are in construction abroad:

- The Soyuz pad, in French Guyana, at the European Space Agency's base.
- Alcântara, a Tsyklon 4 project between Brazil and the Ukraine.

There are two main recovery zones: Arkalyk in Kazakhstan and Orenburg in south-ern Russia and these are also examined. The chapter concludes with an examination of other ground facilities, including Star Town, mission control and the tracking system.

BAIKONOUR

The best known cosmodrome is Baikonour. Legally speaking, it is Russian territory inside the Republic of Kazakhstan. It is probably the largest cosmodrome in the world, 90 km from east to west, 75 km from north to south, with an area of $6,717 \text{ km}^2$ and a downrange fall zone of $104,305 \text{ km}^2$, making it as large as some small countries.

Baikonour cosmodrome is located on the endless, flat and arid desert of Kazakhstan, on the rail line along the Syr Darya between Moscow and Tashkent. In fact, the real Baikonour was a sleepy railhead 350 km far to the north, but the USSR called the cosmodrome Baikonour in the hope that the duped Americans, should they ever attack, would target the hapless citizens and railway workers of the railhead named Baikonour in error. Most of the workers live in the city of Baikonour, formerly Leninsk and before that Tyuratam, between the river and the railway and adjacent to Krainy airfield to the southwest of the cosmodrome. In the post-communist renaming of formerly Soviet cities, Leninsk was formally renamed Baikonour in December 1995, creating a new problem which went unconsidered at the time: two Baikonours! Baikonour cosmodrome employs as many as 35,000 people with 95,000 dependants. There is a long commuting distance from Baikonour city to the launch pads and processing areas, at least 20 km.

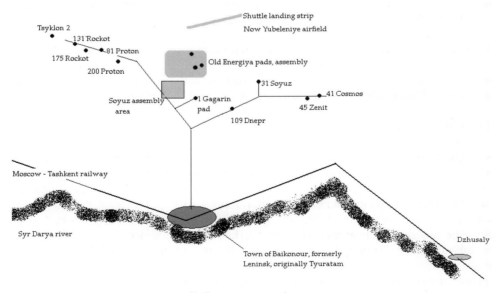

Baikonour cosmodrome

Diesel pulling rocket to the pad

It was not only its location that made it startlingly different from its American rival at Cape Canaveral. The first difference is that rail is the principal mode of transport, both for rockets and the workers of the cosmodrome. The cosmodrome has 470 km of track, although there is a longer, bumpy and poorly maintained road network (1,281 km). Rockets are normally carried flat on their back on railcars from the assembly hangars to the pad where they are then raised to a vertical position. It is a system made easier by the fact that the gauge of the Russian railway is the widest in the world. By contrast, the Americans bring their rockets to the pad on giant road crawlers. Baikonour also has extremely fast turnaround times: it is not unknown for a Zenit to be brought down to the pad, fueled up and fired in a space of less than five hours.

Baikonour cosmodrome was approved 2nd February 1955 and soon construction workers were sent there to build the rocket pad that eventually launched Sputnik. Baikonour was chosen from three possible sites, the others being another Kazakh desert site and Marisky on the western shore of the Caspian Sea. A key determinant was the impact zone intended for the R-7 rocket, the launch site being calculated on the basis of the distance back for a sub-orbital mission. Although the launch site was intended for testing military missiles, the military worried that Baikonour was uncomfortably close to the USSR's southern border and potential enemies. Korolev, looking for a site as close to the equator as possible to get large payloads into orbit, got his way as usual, a clear indication that space exploration was his true priority [1]. The military eventually got their missile base in Plesetsk, but later.

Gagarinsky Start

Unlike the old American launch pads at Cape Canaveral, which were allowed to rust as new gantries were built farther up the Cape, this original pad is still in use. From pad 1 rose Sputnik, the first Lunas, Vostok, Voskhod, most Soyuz and Progress spacecraft. It was renamed "the Gagarin pad" (in Russian *Gagarinsky Start*) and crews for the International Space Station still leave Earth from there. A low assembly building, called MIK, was built near pad 1 in 1956, greatly extended in 1975 for the Apollo Soyuz Test Project and used until the late 1990s when new facilities were built, with a fresh railway connection.

A launch from the Gagarin pad is quite unlike a mission from Cape Canaveral. The Soyuz rocket, 50 m long, clunks and trundles its way down to the pad on a railway flatcar at walking speed, accompanied along its overgrown verges by scores of rocket workers and sightseeers. The transport reaches the pad which is a concrete platform on heavy cement legs. Around it is a giant flame trench, looking like a reservoir empty of water. Once at the pad, the transporter arm lifts the rocket upward. The booster is held up to the vertical while clamps rise up like a bear trap to grasp it so that engineers may inspect it at all levels. An hour before liftoff the high gantries are lowered. New fuel still has to be pumped on board till the very end. Liquid oxygen boils at $-190\,°C$ and wisps of it always surround a rocket's stages. The fuel hoses are pulled away at 60 sec before liftoff. With 20 sec to go, the electrical lines are removed. The rocket's electrical systems must use their own batteries now. The ignition command is sent and flames roar out into the trenches. When the thrust exceeds the rocket's own weight, the four lower arms still restraining it fall back like mechanical petals on a flower and let the rocket go free. Indeed, the system is called "the tulip" (in Russian *Tulipan*).

The Gagarin pad is still very much in use. Money was found to have the pad completely repainted, mainly in military olive green but some parts in orange. A star

Soyuz on the way to the pad on a misty morning

is painted on a steel strut on the side for every launch, over a thousand at this stage, including missile tests. Just north of the pad is a small underground firing bunker which one enters through a metal door. The layout and instrument panels are still much as they were in the 1950s, including the firing key for ignition, with two periscopes for the launch controllers to see the rocket. Chief designer Sergei Korolev used to direct launches from here.

The Gagarin pad became so busy that in 1964 work began to convert a second pad for manned launches. This pad actually dated to the early days of Baikonour and had been used since January 1961 for the missile version of the R-7, part of the Soviet Union's first strike force. This was called pad 31, a distant 20 km to the east, and this subsequently became the base for the launching of many Soyuz, Progress, Zenit and Yantar military missions. It has its own assembly building, called the MIK-40, where Soyuz rockets are assembled. Over 350 launches have been made from there. When a group of German rocket enthusiasts visited there, they had expected to see the crude craftsmanship that supposedly characterized the assembly of Russian rockets. Not a bit, they said: it was full of shining steel rocket bodies and "each single rivet was perfect" [2]. Pad 31 also serves for the new Soyuz 2 rocket.

At around the same time, about 35 km northwest of the original Sputnik pad, construction began of pads to support the UR-500K Proton booster built for the man-around-the-Moon program. Four Proton pads were built in two neighboring complexes, called areas 81 (military) and 200 (civilian). Each pad was flanked by two 110 m high towers which combined the functions of lightning conductor, TV camera point and floodlamp location. The first pads, launch complex 81, were built in the 1960s and served until 1988 when a refurbishment program began. A second set of pads, launch complex 200, was built in the 1970s only 600 m away. This was closed for the year in 2001 for a restoration program. These double Proton pads are all some distance from Baikonour town—an hour on the train. The pads do not have the huge quarry-like trench characteristic of the Gagarin pad: instead, there is a four-star-shaped set of clamps with a number of small flame trenches. The 50 m tall launch

Proton in assembly

tower still includes the room at the top through which cosmonauts were to have been installed in their Zond cabin for a flight to the Moon—it was never removed. The tower is rolled back 5 hr before launch. One of the 200 pads is now being modified for the Angara/Baiterek launcher.

The assembled rocket is moved by diesel railcar to the pad about five days before liftoff. Because the Proton uses storable fuels, there are no telltale signs of valving cool fuels to herald an imminent launch. Like the Gagarin pad, the Proton has its own underground bunker control room, little changed from the 1960s. At launch, there is a single, dull thud, Proton lifting off in 2.2 sec and clearing the tower in 6 sec. Riding a pillar of blue flame, Proton pitches over in 18 sec. A sonic boom is heard a minute into flight. On a clear day, Proton may be followed 5 min into the flight to second- and even third-stage ignition, leaving a wispy contrail behind in the sky.

Protons are assembled in a 120 m long, 50 m wide low horizontal assembly and integration facility which can hold up to six full Protons at a time. The rockets arrive there by rail directly from the Khrunichev factory in Moscow and once there the payload is fitted, the stages are integrated, pressurized and tested. These assembly buildings, completed in 1981, are some of the most modern at Baikonour and run by Khrunichev, the company which operates the Proton. There are even two adjacent hotels to host the Khrunichev and Western workers who accompany a satellite on the final month of its launch campaign, as well as apartment blocks for the workers permanently assigned there (locally called "Proton City"). The first new construction there in several years was assembly building 50, designed to handle the Briz M upper stage, the Proton M and Western satellites. Although the clean rooms are the size of large aircraft hangars, the entire air is evacuated and replaced every eight minutes. Like the other rockets at Baikonour, Protons are brought down to the two pads 6 km and 10 km away clasped on the back of railcars.

In the next wave of expansion, in the mid to late 1960s, construction began of the N pads, set up to support the man-on-the-Moon effort. They are located a mere

Proton launch

3.5 km from the first R-7 pad. The completed N complex was impressive: the two enormous pads, matched by 183 m tall towers, were fed rockets from a 250 m long assembly hangar, about the size of the old Zeppelin airship sheds. After cancelation of the N-1, they lay idle for several years, until in the 1980s engineers began reconstructing them for the N booster's replacement—Energiya. Huge, white-painted facilities were built to support Energiya and the Buran space shuttle it was designed to fly: a Buran integration hall, an Energiya integration hall, test facilities, two parallel launch pads for Energiya–Buran (built on the exact site of the N-1 pads), one of which was used to launch Buran in 1988. A third, quite different but adjacent pad was used to fly Energiya on its first, Polyus SK1F-DM, mission in 1987. At one stage, 4,000 technicians worked here, but most of the facilities fell into disuse when the Buran program was canceled in 1993. The N-1/Energiya pads are no longer safe,

Buran at the pad

nor are the hangars where old Energiya equipment is kept. The two giant Energiya transporters are still there, astride the eight railway lines they used to bring Energiya down to the pad.

In summer 1995 the decision was taken to modify some of the Energiya–Buran facilities and bays for use as integration halls for Western commercial payloads and equipment for the International Space Station. A smaller building was put up alongside the large halls and this survived the collapse of the three large bays in the main building in 2002. Soyuz and Progress spacecraft en route to the International Space Station are now kitted out there as well and it is now the core of modernized Baikonour. One bay is used by the Starsem company, the French–Russian company which operates commercial Soyuz payloads. There is a now a direct railway line from there to pad 1.

To the north of these pads and east of the Proton pads was built the runway for the space shuttle Buran. Called Anniversary airfield (*Yubeleniye*), it is 4,500 m long and 84 m wide and ran from southwest to northeast. The *Yubeleniye* runway was made with polished high-grade concrete to a standard that permitted a surface variation of no more than 2 mm every 3 m. It was only used once for the purpose for which it was built, when on 15th November 1988 the Buran came into land there, the automatic navigation system bringing it in less than 1 m from the designated touchdown point and staying less than 50 cm from the centerline until wheelstop [3].

In the early 1990s, the runway cracked and deteriorated. In 1995, repair work began on the runway which was designated the principal airfield receiving components of the International Space Station from Europe and North America and was re-paved. Boeing 747 and Airbus airliners now fly in there, normally at least one a day in support of upcoming space missions.

Two pads were built in the 1980s for the Zenit launcher, to the south of the second R-7 launch site, pad 31. One of these was destroyed in the launch accident of 1990. It was never repaired, presumably because of cost and because it was presumed that Zenit launch rates would be low. The other pad, though, is kept in immaculate condition. Zenit has its own assembly buildings, jointly operated by the cosmodrome

Buran after landing

construction company, KBTM, with the Yuzhnoye Design Bureau. Due to the slow rate of Zenit launches, only half the facilities are actively used, but this could change with the promised development of a land-based version of Zenit 3SL Sea Launch, called Land Launch. Thirty-six launches have been made from there.

In addition, a number of pads serve for military or military-related launch vehicles, such as Dnepr, Rockot and Strela. There used to be two pad 90 Tsyklon 2/M pads, which between them have seen 105 launches, but one was taken out of service in 1988. The other continues to be used for Tsyklon 2 and may be used for its replacement, the Tsyklon 2K.

Baikonour's launch pads

1	Soyuz (Gagarin pad)
31	Soyuz, Soyuz 2
45	Zenit (double, but one disused)
81, 200	Proton (double) (including one for Angara/Baiterek)
90	Tsyklon 2/M
109	Dnepr
131, 175	Rockot
132	Strela
(41	Cosmos 3M)

In addition, there are pads 41 (Cosmos 1, 3) and 110 (Energiya) which are disused. Baikonour played an important part in the testing program for the military. Pad 41 was used for the development of the R-36 missile, and it was here in October 1960 where one exploded, causing the world's worst launch disaster, killing the chief of the Soviet missile forces, Marshal Nedelin and ninety others.

Neither Russia nor Kazakhstan were sure how best to manage Baikonour when the Soviet Union broke up in late 1991. Kazakhstan had declared its independence during the coup in the Soviet Union and in September 1991 new Kazakh president Nasultan Nasurbayev announced that it was taking over the cosmodrome, except for the military facilities there.

In the course of 1992–93, Baikonour was a kind of no man's land. Technically, it was run by an interstate commission. Workers on the ground made jokes about who owned the table in the room: Russia? Kazakhstan? Both? Half one side and half the other? The Russians had already gone to some effort to keep Kazakhstan on side. In 1991, they had invited Kazakhstan (then a state inside the Soviet Union) to fly a cosmonaut to Mir. Two candidates—Toktar Aubakirov and Talgat Musabayev—were sent to join the cosmonaut squad and Aubakirov flew on Soyuz TM-13 (he later became Deputy Minister of Defence and member of the Baikonour interstate commission). Musabayev stayed on within the cosmonaut corps with the aspiration of making a mission himself, which he eventually did. The Russians probably hoped that with these kinds of arrangements it would be possible to maintain the operation of Baikonour for the foreseeable future. The Russians took the view that they had paid for, built and continued to maintain Baikonour and they did not feel that they owed anything to the Kazakhs.

Baikonour in winter

In fact, joint operation did not work as easily as the Russians hoped. There was much awkwardness as to who actually gave the orders there. Whenever Kazakhs tried to assert their authority, Russians responded by saying that they were part of a military unit, answerable to Russian military law (technically they may have been right). The Kazakhs established a customs post in Baikonour to decide what could move in and out of the cosmodrome, to considerable Russian annoyance.

In July 1993, Russian Defence Minister Pavel Grachev flew to Baikonour for talks with his Kazakh counterpart, Sagadat Nurmagambetov, in what turned out to be a fruitless effort to resolve the issue. The question of the legal authority in the site was indeed a crux issue, but not the most decisive one: money was. Later in 1993, Kazakhstan made it plain that it intended to charge Russia for the use of the cosmodrome and for recovering crews from orbit. Things came to a head in January 1994 when Kazakhstan began to charge Russia prohibitive rates for basing its recovery helicopters in Kazakhstan during the return of the Soyuz TM-17 crew of Vasili Tsibliev and Alexander Serebrov. The capsule was even impounded by customs on landing.

As a result, Russia based its helicopters in Chelyabinsk on Russian territory, flying them into Kazakhstan only for the immediate period of the recovery itself. Even then they had to file flight plans in advance and carry parachute bags full of cash for fueling stops. Even recovery Mil helicopters would be boarded, the commanders being required to pay $400 cash on the spot in landing fees. The Kazakh government also introduced a new element into the equation: pollution. It was certainly true that the desert downrange of Baikonour was littered with the debris of impacting rocket stages. In 1993–94 the Kazakh government carried out an ecological survey, instancing toxic fuels in the ground soil, rusting rocket bodies polluting the land and sewerage discharges from Leninsk. However, Russia regarded the Kazakh exercise as less to do with environmental concerns than the subsequent large compensation claims that followed in their wake.

Landing in the steppe

The Russians responded to the Kazakh threat in several ways. First, they considered how to transfer as many launches as they could to the Plesetsk cosmodrome in northern Russia. They also began to recover capsules in Russia, in preference to Kazakhstan, when they could. The first spacecraft to be landed in Russia was the Raduga capsule of Progress M-18, which came down in the Russian steppe on 4 July 1993, using a landing area between Samara in the west, Omsk in the east, skirting around the southern tip of the Ural mountains near Orenburg and Orsk. Second, they began to cast around for an alternative launch site to Baikonour, but within Russian territory. Third, they began to work out a more permanent arrangement for the use of the Baikonour cosmodrome.

It was not feasible to move all space operations to Plesetsk. The manned space station and flights to geosynchronous orbit required the more southerly latitude offered by Baikonour. Almost all the lucrative commercial flights were of comsats to 24-hr equatorial orbit, and these simply could not be reached from Plesetsk. Once the world's busiest space port, Plesetsk's launch rate was actually falling as the military program contracted. The new commercial business was going to Baikonour. The cosmodrome in Baikonour was the only one with Proton and Zenit rocket pads and Russia needed the use of these pads for the foreseeable future.

Negotiations between Kazakhstan and Russia rumbled on. Eventually, in March 1994, agreement was reached whereby Russia would take a lease on the cosmodrome till 2024 (to be more precise, until 2014, with a 10-year option on extension). The area

Baikonour, the former city of Leninsk

of the cosmodrome would be sovereign Russian territory, under the command of Russian troops. Russia had hoped to strike a bargain for the use of the cosmodrome, offering Kazakhstan access to its space program, sweetened by seats on missions to Mir (Soyuz TM-13, 19). In the end, the Kazakh fee was $115m a year, backdated to 1991, payable in cash in hard currency and they took up an outstanding offer of a flight to Mir in any case. No sooner was the ink dry than the Kazakhs explained that this was just the basic rental and that there would have to be fees on top to actually use the site! Russia would have none of this, the Kazakhs blinked first and there were no fees.

In no time, Russian parliamentarians were complaining that the rental was using up half the manned space program budget. The rental was a running sore for Russia and, granted that its space program was virtually running on empty anyway, this was money it could ill afford. On the positive side, Russia could now operate the cosmodrome without interference, even if landings were another matter. The Kazakhs, for their part, argued that some missions flying out of Baikonour were extremely profitable, namely the Proton commercial missions and that they derived no direct benefit from these profits. Few Kazakhs actually worked there: the workforce was mainly Russian.

Russian hopes that this deal marked the end of the matter for the time being were not realized. Relationships between Russia and Kazakhstan worsened in summer 1999, especially when a Proton crashed on 5th July, showering wreckage downrange. Kazakhstan demanded Russia clean up the mess, including any toxic fuels that had fallen on the ground, especially nitric acid, or heptil as it was more generally called. As an indication of its seriousness, Kazakhstan banned the launch of the waiting

Progress M-42 freighter to Mir, even though it used a different fuel. The ban was not lifted until Russia checked for heptil, cleaned up the area affected, paid compensation to those in the debris field, made a cash advance of $50m to Kazakhstan and agreed to come up with a further $65m early in the new year to settle outstanding rental and other dues. Russia had to pay a further round of compensation, €400,000 when the next Proton went down that October.

Despite this difficult turn of events, the Russian–Kazakh relationship calmed down. At the 2000 meeting of the intergovernmental commission managing the cosmodrome, the biggest issue was whether phone calls from the cosmodrome to surrounding Kazakhstan should be charged at the international or the local rate. Although it probably galled the Russians to do so, they now paid their rent on time.

On 16th January 2004, Russia and Kazakhstan signed a new agreement to supercede prior agreements, even though they had not yet expired. The Kazakhs sought an increase in the annual Russian rental from $115 to $200m, but the Russians turned this down flat. What Russia did offer was $100m in assistance to the Kazakh communications satellite program and a role in the construction of an Angara pad in Baikonour. An agreement for the development of this new pad was signed in Moscow in December 2004. This created a joint enterprise called Baiterek, which would build a pad and associated facilities for the Angara rocket by 2009. President Vladimir Putin and President Nasultan Nasurbayev visited the cosmodrome together in June 2006 to watch the launch, by a Proton, of Kazakhstan's first communications satellite, Kazsat. By all accounts it was a friendly event and the Kazakhs were so pleased they ordered a second satellite.

Baikonour presents a picture of contrasts. Although some parts of the cosmodrome have rusted and fallen into disuse, there is a small core of modern buildings where work is concentrated on the International Space Station, commercial operations and the military program. The city of Baikonour comprises a mixture of abandonment and industry, neglect and modernity, with a thriving market for clothing and food, where people still travel on camels. Life goes on in the world's first spaceport.

Since the start of international collaboration on the Mir space station and then the International Space Station, Americans, Europeans and other nationalities have been frequent visitors to Baikonour, joined in their turn by the hundreds of engineers associated with commercial satellite missions. Americans are often taken aback by the way Baikonour operates. Some aspects of Baikonour could not be more different from Cape Canaveral, where a shuttle seems to take all day to reach its pad on a slowly moving crawler and then spends weeks, sometimes longer, on its pad as final preparations are made. By contrast, the Soyuz is pulled to its pad on a clattering railway line by a briskly-moving diesel engine and is erected on its pad within hours, ready to go. If the press corps can keep up on foot, it is doing well.

In Cape Canaveral, shuttle crews are bundled into their orbiter by a small pad crew, with all visitors kept at least 5 km away. At Baikonour, hundreds of people gather around the launch pad, with the fully fueled frothing rocket behind them, machinery humming to keep the fuel and oxidizer at the lowest possible temperature. The three cosmonauts step down from their minibus, marching forward to three

Disused pads at Baikonour

small painted squares on the cement, where they salute and report that they are ready for flight. Officials, friends, pad workers and well-wishers swarm around them to hug them and say *bon voyage*. A friend of departing astronaut Leroy Chiao contrasted the riotous and energetic celebration of Baikonour with the relative sterility of Cape Canaveral [4].

The other big difference is of course the weather. Shuttle missions can only fly in good weather, inside strict temperature limits (well above freezing), with good visibility and are frequently delayed by thunderstorms and wind. Soyuz flies in all weather, from the scorching heat of central Asia's summer, to the glacial winter temperatures of down to −30°C. Neither does fog nor rain seem to make much of a difference, for the launch goes ahead anyway. One cosmonaut recalled being launched in a wind, because he could feel the rocking of the cabin back and forth. The system has been so well refined that it nearly always goes like clockwork. The last time a manned launch of a Soyuz was postponed, once the countdown had actually started, was in 1971. Many years ago, a group of Indian scientists traveled all the way to Baikonour to see the Russians launch their rocket. Thick fog surrounded the site and they presumed the launch had been called off. Rather to their surprise, it wasn't: later it was explained to them that the dim orange glow that they thought they saw in the far distance was actually their satellite taking off.

Despite the long-term agreement between Russia and Kazakhstan, the Russian Defence Ministry decided in 2001, with President Putin's approval, that it would gradually move all military launches from Baikonour to Plesetsk. The decisive factor appears to have been the decision by Kazakhstan to ban Proton flights in 1999 following two launch failures, the ban not being lifted until compensation was made. The military took the view that their space operations could not be dependent on the goodwill of another government. In effect, Plesetsk would be Russia's military space center, with some civilian launches, while Baikonour would be a civil base run by the various enterprises involved (Khrunichev, Energiya, etc.), although some military launches would continue from there. In August 2006, the head of Russian space troops, Vladimir Popovkin, announced the withdrawal of a further 4,000 soldiers from Baikonour and a €100m rehousing program for them if they stayed in the rocket forces at other locations (presumably Plesetsk). By the end of 2006, the rocket forces had transferred the main facilities (e.g., the Proton area) to civil companies and only a few hundred soldiers were left to guard the communications unit, the old Krainy airfield and some military silo pads.

PLESETSK

Plesetsk was historically the busiest spaceport, not just in the Soviet Union but in the world and at the turn of the century accounted for 38% of all the launches ever made. In contrast to Baikonour, only a few Westerners have been there, generally in connection with the small number of scientific missions operated from there.

Plesetsk was the Soviet Union's original missile base, home to its fleet of four R-7 missiles targeted on the United States from 1960 onward, with a strong shield of surface-to-air missiles. Approval for the construction of Plesetsk as a missile base was given by Khrushchev on 11th January 1957 with the codename Angara. Plesetsk went on duty as a R-7 missile base two years later.

Plesetsk

Use of Baikonour in the manned and lunar programs was so intensive that on 16th September 1960 the government took the decision to develop Plesetsk as a satellite launch base. An area $200\,km^2$ was cleared around the town of Mirny, although the site was named Plesetsk, which was actually the name of both the railway station and a village 4 km away. Like Baikonour, names were never what they seemed.

Using one of the military pads, Cosmos 112 was the first orbital launch from Plesetsk in 1966, an event noticed first by the boys of Kettering Grammar School in England who claimed the credit for the identification of the base (American intelligence knew from the beginning, but they didn't want anyone to know that they knew). The Soviet press was not permitted to acknowledge its existence until 1983 and it was not even marked on official maps until recently. Officially, it was just a "military test site" until it was formally renamed a cosmodrome in the late 1990s. The cosmodrome is 200 km south of Archangel. The airfield has a 2,600 m long runway and takes medium-size airliners and transports, most space equipment coming in on an Ilyushin 76. The cosmodrome itself is located 15 km to the northeast of the towns of Mirny and Kochmas on the banks of the Emtsa river and is surrounded by forest. The area comprises a mixture of dense forest, swamp and rocky outcrops, the original clearance teams living in railway cars and tents.

Most of the cosmodrome's workers live in Mirny, which houses up to 80,000 people in nine-floor apartments. Beside a lily-covered lake in Mirny lies the memorial

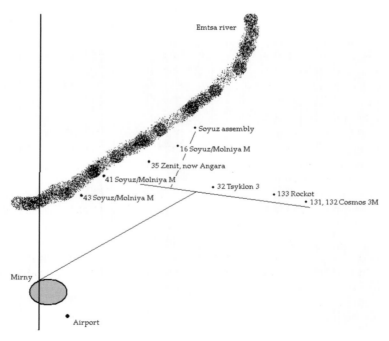

Plesetsk cosmodrome

Mirny

to 51 rocket workers who died in a launch explosion on 18th March 1980, a day always commemorated there and a day on which, by custom, no launches ever take place. The area of the cosmodrome is $1,762 \, m^2$, or 46 km from north to south and 82 km from east to west. Plesetsk is at 63°N, near the Arctic circle. The summer nights are short and it never really gets dark. Precipitation—rain and snow—is about 400 mm each year. In winter there are only a few hours of grayness at mid-day amidst remorseless night. The temperatures are even more extreme than Baikonour, reaching down as low as −46°C, routinely around −20°C in midwinter, not that this has ever affected launchings. Plesetsk is near enough to Sweden for observers there to see the occasional launch in the eastern sky arcing into the far distance. Mirny is about 36 km from the launch pads.

Although it covers a large land area, the core of Plesetsk is actually much more compact than either Baikonour or Kapustin Yar. Because it is built in forestry and in ravines, there is less space to spread the facilities. People go to work on a morning diesel train which leaves Mirny for all the different sites, although some staff are sufficiently close that they can bicycle to their stations.

The present cosmodrome comprises eight pads: two Cosmos 3M, four Soyuz–Molniya, one Angara and a Rockot pad. Following the launch of the last Tsyklon 3 in 2001, the double pad 32 went out of service. Plesetsk has the largest oxygen and nitrogen plant in Europe and has seven assembly shops and integration halls. The road network was in such an atrocious condition by 1992 that it nearly broke the chassis of Boris Yeltsin's limousine when he visited the center. A presidential decree soon led to improvements.

As the military space program contracted, Plesetsk's launch rate declined from one every two weeks to one every two months. In 1994, Baikonour overtook Plesetsk as Russia's busiest launch center, even though Plesetsk had launched so many satellites in the 1970s and 1980s that it would still head the list for some time to

Cosmos 3M at Plesetsk

Satellite arrives at Plesetsk

come (a total of 1,500 launches by 2000, compared with Baikonour's modest 1,100). The fall in launch rate in Plesetsk was gradual and the northern cosmodrome experienced no sudden exodus paralleling the collapse of the Buran/Energiya project in 1993.

Conditions deteriorated in Plesetsk in the mid-1990s. Living quarters for some of the rocket troops were very poor, which must be no joke in the Arctic. One of the city's five schools was made of prefabs. Over two-thirds of the soldiers had to supplement their food by growing potatoes in the grounds of the general hospital, right beside President Yeltsin's quarters during his visit. For the conscripts working there, conditions were harsh, with run-down buildings and food limited largely to bread, gruel, soup and eggs.

How to train the rocket troops was an on-going problem. Each launch required about 300 manual operations to be completed properly, from the erection of the rocket to its correct fueling and error could doom a whole mission. Plesetsk once had the benefit of a training rocket, but the last one was sent to Samara to become a monument. Eventually, in 2001 a basic simulator arrived.

Plesetsk's launch pads

16, 41, 43 (2)	Soyuz/Molniya M
35	Zenit, now Angara
131, 132	Cosmos
133	Rockot
(32 (2)	Tsyklon 3)

The pads are grouped closely together, amid assembly and processing areas. Leaving aside the road improvements resulting from the damage to Boris Yeltsin's limo, Plesetsk began to see the first signs of renewal in the mid-1990s. With investment money supplied by the German partners of its owners, a new launch tower was built for the Rockot launcher, using an old Cosmos 3M site (hitherto, Rockot was launched from silos, but this was unsuitable for civilian payloads).

The Zenit rocket was originally designed to be fired from Plesetsk, but part of the Glushko–Utkin deal for the merging of the Zenit and Energiya program in 1975 was that Zenit have its own facilities in Baikonour. These were built first and construction of a Zenit pad, #35, began in Plesetsk in 1986, but was not completed. When the problems arose with Kazakhstan about the use of Baikonour, the idea of a Zenit pad in Plesetsk was revisited. In the end, the proposal was dropped and the decision taken in 1999 to make the new pad for the forthcoming Angara launcher instead (see Chapter 5). Slow progress was made in the construction of the Angara pad. Although initial construction began, the next set of contractors' supplies did not arrive and work ground to a halt, although money was continually re-allocated to the project. It never seemed to arrive, and so long as this was the case the new site was idle. In October 2005 the Zvezdochka machinery plant in Severodvinsk, a specialist in nuclear submarine repair, completed and handed over the metal structure of the launch clamps for the pad.

(a)

(b)

Integration at Plesetsk. (a) Mission control center. (b) One of the pads. (c) Clean room for payload installation

(c)

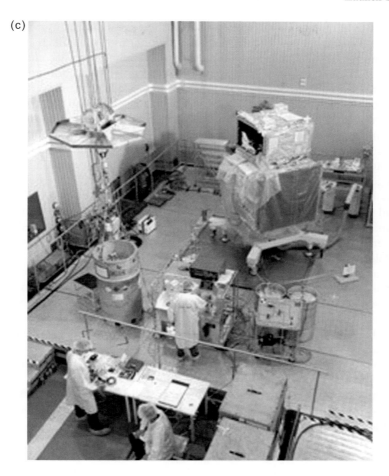

The ecological complaints made about the operation of Baikonour have not been absent in the case of Plesetsk either. At a conference held in Archangelsk on 20th January 2005, green activists complained that the drop zone of 3 m ha had been hit by 18,000 tonnes of scrap metal, 744 tonnes of oxidizer, 652 tonnes of kerosene and 340 tonnes of heptil. The cosmodrome took $10 \, m^3$ of water from the ground every year without paying anything.

SVOBODNY–BLAGOVESHENSK

The idea of developing a new cosmodrome emerged as the difficulties over using Baikonour grew. In November 1993 the Council of Ministers of the Russian Federation issued a decree for a feasibility study of a new cosmodrome, should Baikonour no longer be available. Three new sites were considered, all in the far east:

Vladivostok, Kharbarovsk and Svobodny. The choice was Svobodny, which already had an intercontinental ballistic missile base—Svobodny 18—built in 1968 with 30 underground rocket silos, although all but five of these had been decommissioned in 1993. It is 400 km west of the Sea of Japan and 96 km from the border with China. The nearest town is Blagoveshensk and the two names are sometimes used together.

Svobodny offered a number of advantages. It was close to the trans-Siberian railway—important granted that all Russia's rockets were transported from Moscow by rail—and nearer to the Pacific rim economies, whence some satellite launching business might be hoped for. At 51°N, it was not much farther north than Baikonour, which meant that it was still a relatively economic location for reaching 24-hr orbit. Downrange tracking facilities must be exceptionally good, since rockets leaving Svobodny arch over Sakhalin Island, one of the most radar-intensive zones on our planet and the site of the notorious 1983 incident in which a Korean passenger jet was shot down. There was some local opposition from green activists, worried about the environmental consequences of spilt rocket fuel and falling upper stages on the region. Political interests lobbied strongly for the idea of a cosmodrome as a means to regenerate the economy of the region.

Official authorization for the conversion of Svobodny 18 into a cosmodrome was given by the Russian government in March 1994. When agreement was reached over Baikonour later that same month, there were reports that the Svobodny project would be abandoned, but in the event it went ahead anyway. President Yeltsin visited Blagoveshensk in summer 1994 and the modification of the launch pads in Svobodny began soon thereafter, principally the updating of the old Rockot pads. The President issued a decree, formally establishing Svobodny as a cosmodrome, in 1996. New power supplies and a command center were installed that autumn. By the new century, Svobodny comprised a Rockot-launching area, START launch pad, an industrial area, a fueling plant and an airport. A development plan for the cosmodrome envisaged a technical staff of 30,000 and an eventual total population of 100,000.

Although the authorization for the cosmodrome provided for the construction of Soyuz, Strela and Angara launch pads, the first launch from Svobodny was more modest, using a former military START 1 rocket. The inaugural launch duly took place on 4 March 1997, placing in 400-km orbit a small, 87-kg Strela-class military communications satellite called Zeya (named after the local river). The apprehensions of the environmentalists were borne out when the second stage impacted on Keptin, Yakutia, 35 km downrange, sparking local protests. Later that year, a small American imaging satellite called Early Bird was launched from Svobodny, the first commercial launch, to be followed by others from Israel.

The Russian Space Agency had never been enthusiastic about Svobodny, preferring to concentrate scarce resources on existing facilities. With the settling down of relationships with Kazakhstan over Baikonour, the need for an alternative set of launch sites diminished. In 2005 there were confused reports about the future of Svobodny, some saying that it would be closed by 2009. The military indicated that they would not use it and would focus entirely on Plesetsk, but still this left the door open to commercial launches from there.

First launch from Dombarovska, 2006

DOMBAROVSKA/YASNY

Like Svobodny, Dombarovska was also a missile base, home to 45 silos of R-36 missiles from 1973 (Tsyklon is a civilian derivative). It is only 15 km from the Kazakh border and has its own airfield.

The first launch from Dombarovska took place on 12th July 2006, when the Dnepr rocket put into orbit a private American demonstrator spacecraft, Bigelow Aerospace's Genesis 1 inflatable spacecraft, intended to test out the idea of inflatable space stations. The idea actually went back to the 1960s and was popularized in the *Drift Marlo* cartoons, which portrayed an orbiting, ringed, rubber-made, inflatable space station. Once in its planned 555 × 561-km, 64.5° almost circular orbit, the module inflated to its maximum 3.8 m diameter and deployed a set of solar arrays which soon began supplying power, all to the satisfaction of Bigelow's mission control in Las Vegas. Cameras soon relayed back images of the inflated module, both from the outside and from the inside. Bigelow planned a set of further demonstrations, pursuing a line of development that offered the deployment of large, low-cost space station structures, provided of course that their structural integrity can be preserved.

For launches from the *Odyssey* platform in the Pacific Ocean, see Zenit 3SL in Chapter 4 (p. 170–5); and for Barents Sea launches, see Shtil and relatives, Chapter 4 (p. 185–7). For the location of Dombarovska, see p. 238.

SOYUZ À KOUROU, FRENCH GUYANA

The improbable idea of developing a Russian launch base in the European Union and French territory in the South American jungle arose from a confluence of factors in the early 2000s. First, it emerged from the Starsem alliance, a joint European–Russian enterprise for the development of the Soyuz rocket, uniting leading European aviation company EADS with the producer of the R-7 Soyuz rocket in Samara, the company TsSKB Progress. This was a successful undertaking, enabling Western

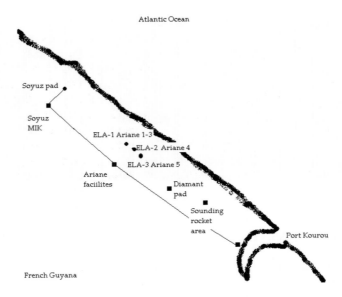

Soyuz à Kourou, Soyuz launch site, Kourou, French Guyana

companies to put communications and other satellites into orbit at competitive rates and bringing TsSKB Progress much-needed investment and modernization. Second, the European Space Agency found in the early 2000s that it had a successful launcher for the heavy end of the commercial launcher market, the Ariane 5, and an up-and-coming launcher for the small end, the Italian–French Vega—but nothing in between [5].

It was the Russians who got the process rolling by making a formal request in spring 2001 for permission to fly the Soyuz from Kourou and it was tabled for discussion during a visit of French research minister Gérard Schwartzenberg to Moscow that April. The proposal was at once referred to a study group in the European Space Agency.

Russian officials paid their first visit to Kourou in 2002 and identified a possible new site to the north of the existing launch pads. The Soyuz-at-Kourou project was approved by the European Space Agency at its May 2003 meeting. Texts were agreed by the Russian Space Agency, on the one hand, and the French government, on the other, in Moscow in October 2003, the French side being represented by Prime Minister Jean-Pierre Raffarin and Research Minister and former Soyuz cosmonaut Claudie Hagneré. During a visit to Paris by President Putin, the agreement was formally signed between Russia, the European Space Agency and the French space agency CNES on 5th November 2003. The first site inspection took place in February 2004. The global satellite market soon indicated its interest by booking a first launch in 2008, the Australian Optus D2 comsat, followed by three more orders.

Several obstacles had to be overcome. First, there were American and European security concerns about the permanent presence of a Russian colony of 200 launch

Kourou—Soyuz jungle take-off

specialists. Would they spy on commercial Western satellites due to be launched? National military observation satellites were also launched from Kourou. Second, there was the problem of funding. The project cost €344m, of which €223m would come from the European Space Agency countries supporting the project: France (63.13%), Germany (5.65%), Italy (8.71%), Spain (3.26%), Belgium (6.53%), Austria (1%) and Switzerland (2.72%) (ESA members are obligated to support the science program, but opt in to other projects, like launchers). There was still a shortfall of €121m, which precipitated a minor crisis. The European Union would not guarantee a loan for this amount. France had rarely failed to back European launcher projects in the past: this was no exception and France stepped in to guarantee a loan to this value from Arianespace, to be paid back from the subsequent profits. This cemented the French lead in the project, legally run by a three-sided treaty between Russia, France and the European Space Agency. ESA is responsible for overall management, CNES is the system architect, Arianespace is the operator, Roscosmos is responsible for Russian management and Lavochkin deals with the upper stage. Late in the day, the European Union eventually contributed a modest €18m.

French Guyana is known to many people through the Dustin Hoffman film *Papillon*. The site was settled by France in the 1760s: most perished in the jungles of the mainland, the survivors fleeing to the three islands off the coast, called the Iles de Salut, or Islands of Salvation. These three islands were Devil's Island, Ile Royale and Ile Saint Joseph and in the 1850s they were turned into notorious penal colonies. The prisons closed in 1947; when the rocketeers arrived in 1964 they found nothing but ruins and jungle.

The launch site is of course on the mainland, though rockets curve out to sea over Devil's Island. It is located barely north of the Equator, 5.14°N, which gives satellites a huge velocity advantage in reaching equatorial orbit. The launch site is near Cayenne, a coastal plain 29 km wide and 60 km long served by a port where the rockets arrive by sea from France. The only disadvantage is its hot and steamy climate: the wet season is very wet. The site covers 1,000 km², about 1% of the land area of the country. Temperatures range from 18° to 34°C, with an average of 26°. Rainfall is 2.9 m a year. Nowadays, 75,000 people live in Guyana. The interior is dense jungle and is used for training by the French Foreign Legion, where it still has a base. Piranhas and alligators infest its rivers. The native Amazonian people still hunt there in traditional ways, but entry into the interior by visitors is discouraged.

France began construction of the space center there in 1964 and it was designated Europe's launching base two years later. The first Ariane flew from there in 1979. Five launching pads were built: one for sounding rockets, one for the small French Diamant launcher, one for Ariane 1 (now in reconstruction for Vega), one for Ariane 4 and a double pad for the Ariane 5. Around them are production and preparation facilities, integration halls, clean rooms and a launch control center. Some 1,400 people live permanently around the launch center.

Soyuz will have its own dedicated area at Kourou, 10 km north of the Ariane pads, called the Soyuz zone. Soyuz will be launched from the new pad called the ELS, *Ensemble de Lancement Soyuz*. The bulldozers began clearing the ELS in November 2004. Construction involves the building of a vehicle assembly building, like the one at Baikonour, called the MIK (integration and test building hall in Russian). This is a low structure, only 20 m high but 56 m long, in which up to two Soyuz can be assembled horizontally at a time, having been shipped from Samara via St. Petersburg. Beside the MIK are technical rooms, a 200-person hardened launch control

Kourou—Soyuz readied for launch

center and storage tanks for the kerosene fuel. The Soyuz is then brought to the pad in the traditional way, on a 1 km long railway line.

The launch pad itself is built according to the blueprints used for the Gagarin pad in Baikonour and involves the similar construction of a launch platform and 26 m deep flame trench. The Soyuz rocket will be clamped by the same type of four-arm tower as at Baikonour, with the familiar system of weights and pullies: when thrust builds up to 480 tonnes, the restraining arms are automatically released. There will also be an enclosed mobile tower 56.5 m high for the vertical installation of the Fregat stage and payload. Vertical installation is not the Russian tradition, but it is the European procedure and ESA insisted on it. The enclosed mobile tower was necessary to protect the payload from humidity and rain. The version of the R-7 launcher used will be the Soyuz 2.1.a and the Soyuz 2.1.b, which will be called the Soyuz ST A and ST B, respectively (the terms ST K, K for Kourou has also appeared).

Construction of what the French call *Soyuz à Kourou* (Soyuz at Kourou) began with the first soil turning in January 2005. The deeper the builders dug, the more granite they found and had to resort to blasting to clear the area. The pit was not completed until autumn 2006. The following year 2007 was set for the installation of equipment, 2008 for testing and the first flights starting soon thereafter.

Operating *Soyuz à Kourou* will require a considerable movement of equipment: fuel from Russia, the Soyuz from Samara, the Fregat from Moscow and the payload from Europe. The Russian equipment will go by ship from St. Petersburg, the French equipment from Le Havre. Fuel will be shipped in from Russia and a fuel farm will be set up in Kourou.

Kourou—clearing the jungle

KAPUSTIN YAR: THE VOLGOGRAD STATION

Kapustin Yar is the oldest of the rocket bases. Korolev and his colleagues had moved there to test the German A-4s in 1947 and it was used subsequently for early Russian rockets, missiles and sounding rockets in the 1950s. The base was developed by the military, as were the subsequent cosmodromes, although a separate command called the Strategic Rocket Forces was not created until December 1959.

All the first satellites, from 1957 to autumn 1961, were launched from Baikonour. The use of Kapustin Yar as a satellite launch base dates to October 1961 and was associated with the introduction of the new R-12 (Cosmos) rocket and the DS series of satellites built by Mikhael Yangel's OKB-586 design bureau, now Yuzhnoye. Yangel made Kaspustin Yar his launch center for his early satellite program, though convenience must have played a part, for Kapustin Yar was much nearer to his factory and design bureau in Dnepropetrovsk. The first launch failed, but he succeeded with Cosmos 1 in March 1962, followed by Cosmos 2, 3, 5 and 6.

Kapustin Yar was used as a satellite launching base for 139 missions from 1962 to 1987, mainly small scientific satellites in the Cosmos series, but also a number of small experimental missions, such as spaceplanes. Then it fell into disuse for several years.

In the mid-1990s, Kapustin Yar became, under international auspices, a base for supervising the decommissioning of old Cold War missiles. Arms inspectors would arrive there from time to time to check that missiles were either blown up or put beyond further use. The first signs that it might be used again for space-related activity came in January–February 1997, when two MR-12 sounding rockets were fired from there under a joint program between the Institute for Dynamics of the Geosphere of the Russian Academy of Sciences, The Johns Hopkins University in Maryland and the respective military forces of Russia and the United States. The sounding rockets released artificial plasma clouds to test for their effects on radio communication. Later that year, President Yeltsin visited the cosmodrome to mark the 50th anniversary of its opening when it first fired captured German A-4s across

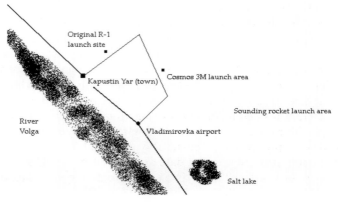

Kapustin Yar cosmodrome (Volgograd Station)

the south Russian desert. He inspected the rocket troops, awarded medals and promised to pay the cosmodrome's debts. The cosmodrome was effectively re-opened briefly in April 1999 for the launching of a German X-ray scientific satellite called Abrixas on a Cosmos 3M. This was the 84th orbital launch.

Kapustin Yar was used only once in the new century, for a Cosmos 3M sub-orbital military test launch on 22nd April 2006. It is unlikely to see more than an occasional launch in the future. Kapustin Yar had two airfields (the old Kapustin Yar, Vladimirovka), three disused R-12 pads, one Cosmos 3M pad (#107), one START pad, tracking facilities and some assembly facilities.

ALCÂNTARA

Brazil's space program dates to the 1960s, when it built a sounding rocket range on the northeastern tip of the country at Barreira do Inferno. In 1979, Brazil decided to build a small indigenous rocket able to put small payloads into Earth orbit, and accordingly developed a new launch range farther north, only 3° south of the equator, at Alcântara near the city of São Luis. The European rocket range in French Guyana lies farther north, on the other side of the equator line. Each site has the advantage of being able to launch straight out to sea over the ocean. Alcântara covers $620\,\text{km}^2$, cost €250m to build and opened in 1990 [6].

After many years, construction of Brazil's rocket, the solid fuel VLS was completed, but the first two attempts to reach orbit in 1997 and 1999 were unsuccessful. Worse was to follow. Brazil was counting down for its third orbital attempt on 22nd August 2003. Hopes were high that success would be achieved at last and two satellites were in the nose. Some hours before the planned launch, an electrical charge ignited one of the stages prematurely: there was a huge explosion, 21 technicians working around the rocket were killed instantly and the pad was destroyed. Russian space specialists were invited to assist with the investigation.

In early 2002, Russian space industry representatives approached the Brazilians who offered their cosmodrome-building expertise to assist the Brazilians in a $2m development of the base. The following year, a treaty was signed between Russia and Brazil at a ceremony attended by both Vladimir Putin and Lula de Silva.

But the Russians were not the only group courted by the Brazilians, for in summer 2002 a memorandum of understanding was agreed between Brazil and Ukraine for the development of the launch center as a joint venture with the prospective Tsyklon 4 rocket, to put 1,800-kg payloads into 24-hr orbit from 2005 for $30m a go. The following summer, Brazil announced that it would repair the launch site and still attempt a satellite launch. Ukrainian officials visited the site in August 2003. Although the project to develop Alcântara appeared to be firmly set in the Yuzhnoye calendar, progress was remarkably slow and the date for a first Tsyklon launch kept slipping, the most recent date announced being 2009.

Finally, proposals surfaced from time to time to develop an equatorial spaceport on Christmas Island in the Pacific. These plans were associated with the Aurora launch vehicle, a derivative of the R-7. Sketches were even done of the 85-ha site

Tsyklon 4

Table 6.1. Launches by active cosmodromes, 2000–6

	Baikonour	Plesetsk	Svobodny	*Odyssey*	Barents Sea	Dombarovska
2000	30	3	1	2		
2001	16	6	1	1	1	
2002	14	9		1	1	
2003	15	6		3		
2004	17	6		3		
2005	19	4		4	2	
2006	16	5	1	5	1	1
All 1957–2006	1,180	1,498	6	21	5	1

on the south of the island with one pad—later four pads—and the full range of service facilities, with an expected employment of up to 550 people. Although backed by the Australian government—a formal agreement was even signed in May 2001—it failed to find private investors, who took the view that the market was already well served by existing rockets, not least Sea Launch which already fired from the Pacific equator. Discussions were also held for other possible spaceports in the region, the most touted locations being Cape York and Papua New Guinea. The 2001 agreement also permitted START launches from the old British launch base at Woomera in the Australian desert. Nothing concrete has yet emerged from these initiatives.

Table 6.1 shows the use of the respective cosmodromes up to 26th December 2006.

RECOVERY ZONES

The principal recovery zone for returning Russian spaceships is around the town of Arkalyk, Kazakhstan, flat steppe land to the north of Baikonour cosmodrome; Arkalyk is still used for manned Soyuz spacecraft returning from the International Space Station. About 200 km north of Baikonour, on the flat steppe, are five towns in a diamond shape: Kustanai, Kokchetav, Tselinograd, Dzhezhkazgan and Arkalyk; Arkalyk is the aiming point [7]. All land in this general area.

Reentry is commanded over the Atlantic Ocean near the coast of Africa, with the Soyuz making a long, sweeping entry over the Arabian desert toward Kazakhstan. Few people seem to have observed Soyuz reentries from the ground, but ISS Science Officer Peggy Whitson once followed Soyuz TM-34 down from her vantage point in the space station. She saw the separation of the orbital, landing and service modules and heard the crew report back *Razdeleniye* (separation). The orbital and service modules sparkled to destruction, while an ever-longer white contrail glowed behind the landing module with its human crew cocooned inside the fireball.

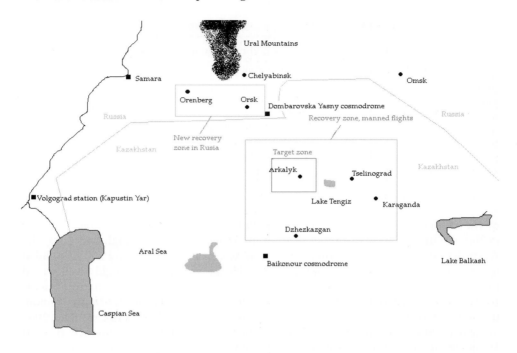

Russian recovery zones, including Dombarovska

Manned spaceraft recoveries 2000–6
Manned

16 Jun 2000	Soyuz TM-30	Sergei Zalotin
		Alexander Kaleri
8 May 2001	Soyuz TM-31	Talgat Musabayev
		Yuri Baturin
		Dennis Tito
31 Oct 2001	Soyuz TM-32	Viktor Afanasayev
		Sergei Kozeyev
		Claudie Hagneré
5 May 2002	Soyuz TM-33	Yuri Gidzenko
		Roberto Vittori
		Mark Shuttleworth
8 Nov 2002	Soyuz TM-34	Sergei Zalotin
		Yuri Lonchakov
		Frank de Winne
4 May 2003	Soyuz TMA-1	Kenneth Bowersox
		Nikolai Budarin
		Donald Pettit
28 Oct 2003	Soyuz TMA-2	Yuri Malenchenko
		Edward Lu
		Pedro Duque

30 Apr 2004	Soyuz TMA-3	Michael Foale
		Alexander Kaleri
		André Kuipers
23 Oct 2004	Soyuz TMA-4	Gennady Padalka
		Michael Finke
		Yuri Shargin
25 Apr 2005	Soyuz TMA-5	Leroy Chiao
		Salizhan Sharipov
		Roberto Vittori
11 Oct 2005	Soyuz TMA-6	Sergei Krikalev
		John Phillips
		Gregory Olsen
30 Mar 2006	Soyuz TMA-7	William McArthur
		Valeri Tokarev
		Marcos Pontes
29 Sep 2006	Soyuz TMA-8	Pavel Vinogradov
		Jeffrey Williams
		Anousheh Ansari

Reentry is a well-practised procedure at this stage. As reentry nears, helicopters take to the air. Electricity lines are turned off as a precaution. The Soyuz is, meantime, firing its engine over the South Atlantic, beginning its long curve that takes it over north Africa and the Middle East. Mark Shuttleworth recalls:

> The actual reentry burn lasts three minutes. The force of the engine firing is innocuous enough, but at the end, you know it will be all over and you'll be on the ground in 35 min. You go into reentry upside down. Then the blackness outside turns red and gravity forces kick in. You feel you have an elephant sitting on your chest and you reach four to five times the force of gravity. For the next two to three minutes you feel that you are inside a furnace. The thrusters rotate the spacecraft, but you can still see the metal melting and the hear glass straining under the heat and pressure. Soon you realize you've broken the back of it. There's a big jerk and a lift when the parachute comes out and you talk to the helicopters on the radio. As you come into land, you brace and remember to put your head back and keep your tongue in. And suddenly it's all over.

For those who have been in orbit for months, the nicest thing is the blast of Earth's fresh air into the cabin, no matter how cold it might be in winter.

That's the perspective from inside the cabin. On the ground, mission control, TsUP, will have followed retrofire and heard from the crew until the moment of blackout. There is little that the controllers can do. On the ground, defence radars try to pick up the falling Soyuz. Mil helicopters are in the air, shepherded by a lead helicopter which follows the Soyuz on its radar until it can see the Soyuz under its parachute. The cameras normally capture the moment when the four solid-fuel rockets fire under the Soyuz at the moment of touchdown, blackening the brown Earth or the white snow underneath.

Helicopters fly out

Soyuz either ends up on its side or in an upright position. Whichever way it is, ground crews landing beside it assist the cosmonauts out. If upright, a slide—just like one from a children's playground–is then placed on the side of the Soyuz and the cosmonauts gently slide down. One after another, they are then placed in folding film director's chairs where they receive flowers and glasses of hot tea. Cameras flash and picture the grinning space travelers while another helicopter crew is erecting a small inflatable field hospital—an arched tent where the cosmonauts can take off their spacesuits and are then given an hour-long medical.

Landing conditions can vary enormously. Some cabins have come down in almost no visibility, others in raging snowstorms. Some have come down in fine weather and the cosmonauts have been greeted by warm sunshine stretching long shadows out on the rough steppe grass, immensely calming after their rough tumble from the sky. Within two hours of landing, the helicopters are back in the air again, heading for the nearest town, be that Arkalyk or Kustanai, where the cosmonauts are welcomed by local officials and Kazakhs in traditional dress offering the standard greeting of bread and salt. By the evening, the returning cosmonauts are back on their way to Moscow, traditionally in one of Star Town's Tupolev airliners where they can relax, eat and sleep off the adrenalin-fueled experience of descent from orbit.

The most problematic recovery was that of Soyuz TMA-1. Normally, the onboard computer guides the Soyuz into a long, shallow, gentle curving trajectory, taking in the continuous measurements of the gyros, making constant adjustments to the path to ensure the correct angle, tilt and rotation for reentry, with the crew able to take over at any time. This time the spacecraft had strayed slightly outside its normal

Helicopter near cabin

angle of orientation just before reentry, but rather than correct it the computer crashed out completely, placing the spacecraft in a steep descent trajectory [8]. The crew experienced 8.5 times the force of gravity (G), compared with the expected smoother 3 G to 4 G. As a result, the cabin came down in spring steppeland 450 km west and short of Arkalyk instead of right beside it, with no one in sight, although there had been some contact between commander Nikolai Budarin and a recovery helicopter during the descent, but nothing since. As the hours passed, there was no word back to Moscow mission control, but since the descent appeared from a distance to be reasonably normal, nobody seemed to worry. Indeed, once TsUP had got the report of parachute deployment, most people had left the control room and only returned when journalists called them to ask what had happened to the crew and why there were no pictures yet of their happy homecoming.

Meantime, the crew members had exited the small cabin unaided, not an easy job; they were sitting back on the spring grass, waiting for something to happen. Eventually, after 2 hr, a helicopter arrived and they were picked up. They were in no danger nor even discomfort, but NASA was unhappy at the manner in which the craft had made an unprogrammed reentry, that its astronauts had ended up far from the planned landing spot, without communications and had spent two hours waiting to be retrieved. Again, the experience highlighted some cultural differences of approach between the two countries. Apart from the computer, which was reprogrammed, the communications problem had a simple solution: from that moment on, each Soyuz was equipped with a mobile phone! A set was duly sent up on the Progress M-48 cargo ship. When Soyuz TMA-2 came down, the Russians tried to reassure their NASA colleagues by putting no fewer than twelve helicopters and three planes into the air for the recovery.

Landing engines ignite

Another problematic recovery was the return of Soyuz TMA-6. There were a number of problems in undocking, as sensors indicated an inadequate seal. This brought echoes of the Soyuz 11 return, in which a valve had opened and the three sleeve-shirted cosmonauts had died. Precautions had been taken against a recurrence, by requiring cosmonauts to wear full spacesuits during the descent. Despite the go-ahead to undock, pressure fell in the returning cabin from the normal 780 mm to 660 mm. Falling pressure had been first noticed just after the de-orbit burn by space tourist Gregory Olsen, who had taken responsibility for pressure and electrical systems. The emergency procedure swung into operation and the Sokol pressure suits fully pressurized themselves with 40% oxygen (they are vented above that level, for fear of fire) [9]. Later investigations suggested that a buckle strap had become caught in the docking seal, prompting a new procedure for more rigorous checks for foreign objects. Later, returning mission commander Sergei Krikalev was to admit that the situation had been "fairly serious".

Generally, landings were timed to take place in the early morning, which meant that the rescuers had an entire day of daylight in which to find the returned cabin. This was not always the case. The landing of Soyuz TMA-4 was delayed due to the late arrival at the ISS of Soyuz TMA-5, due in turn to a number of minor pre-launch delays (a bolt had to be reinstalled). Soyuz TMA-4 came down in darkness, but it was located by helicopters as it parachuted down.

As soon as they landed, the cosmonauts were normally brought into a medical tent. What the doctors noticed most was fatigue, disorientation and imbalance, problems with posture and a loss in bone mineral content. Cosmonauts were routinely sent for two months to a health resort and this invariably eliminated the

Recovery of Yuri Malenchenko and Ed Lu

Flight back to Moscow—Nikolai Budarin, Talgat Musabayev and Ken Bowersox

effects of weightlessness, the exception being bone mineral density, which never fully recovered [10].

Traditionally, military photo-reconaissance satellites were brought back to Earth in the main landing zone in Kazakhstan, slightly on the western edge of the area where Soyuz spacecraft came down. It is understood that, following the dispute with Kazakhstan, the site was moved slightly to the northwest, to come down in similar terrain nearer to the towns of Orenburg and Orsk, over the border in southern Russia [11]. Later manned landings will be moved to Russia, once the Soyuz TMM series is introduced, for it has greater landing accuracy.

Typically, Kobalts (see Chapter 4) orbited for 120 to 130 days, but Cosmos 2410 was brought back early, after 107 days. Something appears to have gone wrong and a number of irregular maneuvers were made in orbit. Whatever happened, the cabin was never found. Initially, it was thought that it was lost in snow, but it is possible that it did not survive reentry. In addition, small film capsules are sent down from Kobalts in orbit, two in the case of Cosmos 2410, and these are recovered in the same area. Kometa topographic and mapping satellites are normally recovered after 44 days.

Unmanned recoveries, 2000–6

Recovery date	Mission	Days on orbit
9 Feb 2000	Fregat IRDT 1	Less than one*
29 Sep 2000	Cosmos 2373 Kometa	47
10 Oct 2001	Cosmos 2377 Kobalt	133
27 Jun 2002	Cosmos 2387 Kobalt	122
12 Jan 2005	Cosmos 2410 Kobalt	107**
15 Jun 2005	Foton M-2	16
15 Oct 2005	Cosmos 2415 Kometa	44
19 Jul 2006	Cosmos 2420 Kobalt M	77

*Part found   **Not found

Explosive de-orbits

Recovery date	Mission	Type	Days on orbit
9 Dec 2003	Cosmos 2399	Don	120
17 Nov 2006	Cosmos 2423	Don	65

DE-ORBIT ZONES

Russia uses a single de-orbit zone to take out of orbit disused spacecraft, and this is in the Southern Ocean east of New Zealand at 40°S, where they burn to destruction away from shipping lanes. This is principally used for the Progress re-supply freighter to the International Space Station, but also for non-recoverable military spacecraft that have reached the end of their lifetimes. Most are taken out of orbit with a single burn, but in the case of Cosmos 2383 the satellite was taken out of its operational

Foton landing in snow

orbit a week earlier, but in such a way that it would quickly fragment. The residual propellants of the Cosmos 2367 US P exploded before it was finally de-orbited. The following list catalogs all Russian spacecraft deliberately taken out of orbit in the new century, but does not catalog those that have decayed from orbit naturally.

Russian spacecraft de-orbited, 2000–6

Date	Spacecraft	Mission
14 Oct 2000	Progress M1-2	Mir
1 Nov 2000	Progress M1-3	ISS
29 Jan 2001	Progress M-43	Mir
9 Feb 2001	Progress M1-4	ISS
23 Mar 2001	Mir	Space station, 1986–2001
19 Apr 2001	Cosmos 2372	Orlets Yenisey (photo-reconnaissance)
3 May 2001	Cosmos 2370	Neman (photo-reconnaissance)
21 Aug 2001	Progress M1-6	ISS
14 Oct 2002	Progress M-44	ISS
22 Nov 2001	Cosmos 2367	US P
28 Aug 2003	Progress M-47	ISS
4 Oct 2003	Progress M1-10	ISS
28 Jan 2004	Progress M-48	ISS
28 Feb 2004	Cosmos 2383	US P
3 Jun 2004	Progress M1-11	ISS
30 Jul 2004	Progress M-49	ISS
23 Dec 2004	Progress M-50	ISS
9 Mar 2005	Progress M-51	ISS
15 Jun 2005	Progress M-52	ISS
7 Sep 2005	Progress M-53	ISS
3 Mar 2006	Progress M-54	ISS
24 Mar 2006	Arabsat 4A	(Result of launch failure)
19 Jun 2006	Progress M-55	ISS
19 Sep 2006	Progress M-56	ISS

OTHER GROUND FACILITIES

The cosmodromes are the most visible and the largest of the ground facilities sustaining the Russian space program. Equally important are other key facilities such as Star Town, mission control and the tracking services.

STAR TOWN, TsPK

Zvezchny Gorodok, Star Town, also called Star City (the word *gorod* in Russian can mean either), now also known as the Yuri Gagarin Cosmonaut Training Center (TsPK in Russian) dates to 1960, when facilities had to be found to train and house the newly-formed cosmonaut squad. Three hundred and ten hectares of land were cleared in a birch forest 40 km northeast of Moscow. It was a closed city until the mid-1990s. Now, visitors can get off at the Tsiolkovsky stop, seventeen stations from Moscow city center, on the line to Monino.

The central point of Star Town is a man-made lake originally built in the early 1970s by conscripts, around which are a series of 15-floor blocks where the cosmonauts, their families and Star Town workers live. The lake freezes over in winter and it

Entry to Star Town in winter

is possible to walk, sledge and ski on the lake. Farther away lie health and sports facilities, a museum, post office, shops, nursery and hotel [12]. Star Town is a complex of offices, apartments, tree-lined avenues and functional buildings, the purpose of which can often be guessed from the outside appearance—for example, the dome-shaped planetarium, where navigation is taught.

The center expanded in three waves. The first was 1969–74, with the addition of a large hall for space stations and a new centrifuge. The second was in 1980, with the building of the hydrolab where cosmonauts would test spacewalks underwater. The third was in 1984, with the addition of facilities for the Buran space shuttle. These included a huge building for a full-scale mockup with robot arm, with a large amount of underground electrical cabling. A simulator was built to assist in training for the Mir space station, called KTOK (literally, "complex for simulators for spaceships"). No new facilities have been built since then, although plans were made. The Buran facilities were abandoned and have been left as shells exposed to the elements where young people sometimes hold outdoor drinks parties and ignore the dangers of the crumbling structures.

Star Town is a place of contrasts. There has been little maintenance in most of it for many years and the offices have a definite 1980s feel. Parts of it stink where the toilets haven't been cleaned for years. Space tourists remarked on how trees were growing through some of the buildings, passages had no light bulbs, but still the place was humming with energy.

The main working facility is a series of twelve blocks comprising full-scale space station training replicas, simulators, centrifuge, running track, administration building, swimming pool and hydrolab. The hydrolab is 23 m across and 12 m deep, holding 5 m litres of water. The centrifuge was built by Swedish engineers in 1980.

Centrifuge in Star Town

Snowy Star Town

Weighing 300 tonnes, it is 18 m long and can fling two trainees around at a time at up to 68 revolutions a minute, treating them to up to 30-G forces.

Star Town became a much more open and international place. Besides the Russian cosmonauts training and living in Star Town, there were contingents of Americans, Japanese, Europeans and occasionally Chinese. The European and American space agencies even set up liaison offices there. Star Town did not suffer the same deterioration as did the cosmodromes, despite little new building and maintenance being problematical (a general problem in Russia in any case). The most visually striking additions were special houses built for the NASA astronauts, which, being built in New England style, seemed incongruous and out of place amid the Soviet functionalism of the surrounding architecture. Cycle stands were put in place for the American astronauts to get around and exercise.

What about the cosmonauts themselves? From 1960 the Soviet Union and then Russia recruited several hundred people in over 30 groups of cosmonauts. Essentially, there are three strands to the cosmonaut squad. Military officers and pilots have always been the dominant, core element of the squad and they comprised the first, historic group of twenty young cosmonauts. Almost all Soyuz missions are commanded by a military pilot, as is the case with the American space shuttle.

The second group were flight engineers. Originally, they were drawn from the OKB-1 Korolev Design Bureau—indeed, it was once called the Korolev kindergarten. Even today, the second largest part of the cosmonaut squad consists of RKK Energiya engineers and their number reflects the continued dominance of Energiya in the Russian space program. As often as not, the engineers are older than their military commanders. These engineers have a scientific interest but also know the ins-and-outs of a space station's systems intimately and are expected to fix things that go wrong. In one sense, the engineer, although junior in rank, may be more crucial to the success of a mission than the commander. The third part of the cosmonaut squad might be termed miscellaneous and consists of doctors drawn from the Institute for Medical and Biological Problems (IMBP), engineers and specialists selected from design bureaus other than Energiya, and others recruited for particular missions.

The early cosmonauts became household names. By the time of the long space station missions to Mir, the glamor had worn off and the role of cosmonaut became seen as a normal, albeit unusually demanding, profession. The selling of seats on missions to other space agencies (and tourists) had a negative effect on the cosmonaut squad, for every seat taken by a foreigner was one less for them. As a result, opportunities for Russian cosmonauts to fly into space diminished. Granted the fact that the position of mission commander normally went to a veteran, the opportunities for a new cosmonaut to fly reduced in number. Between old cosmonauts hanging on for another mission and new recruits, there had always been more cosmonauts in the squad than there would ever be missions for them, so the competition became more intense than ever.

To be accepted for training, prospective cosmonauts had to pass three months of medical tests, with doctors quick to pounce on any irregularity. Medical examinations continue at regular intervals, so one never entirely escapes this aspect of being a cosmonaut. Once accepted, the main form of instruction is through lectures from trainers and designers, followed by regular technical examinations. There are written examinations for each mission, which every crew member has to pass to fly. The system strikes American visitors as old-fashioned and formal, for NASA's main emphasis is on simulator-based training. Despite that, Americans are always impressed as to how well Russian crew members know the various systems and sub-systems on the space station. Generic training lasts two years, after which a cosmonaut will hope to be assigned to a specific mission.

Training for a long-duration mission on the space station is lengthy, about two years and involves a lot of traveling. Cosmonauts and astronauts not only alternate between training in Moscow and Houston, often a month at a time, but they visit the factories where other space station equipment is built, such as in Europe, Japan and Canada. They are away from their families for long periods, following which they are off the planet altogether for six months. Saying goodbye to crews as they fly out to Baikonour and then welcoming them back half a year afterwards are big, emotional events.

Despite the large number of Soviet period cosmonauts still available, TsPK continued to recruit in the 1990s. There would always be a need for younger cosmonauts to take the place of those retiring and, in anticipation of better times, the prospect that more cosmonauts would get to fly. These were as follows (p. 251).

Soyuz training simulator in Star Town

Civilians, 1992
Alexander Lazutkin
Sergei Treshchev
Pavel Vinogradov

Civilians, 1992
Nadezhda Kuzhelnaya
Mikhail Tyurin

Pilot, 1995
N.A. Pushenko

Pilot, 1996
Yuri Shargin

Civilians 1996
Konstantin Kozeyev
Oleg Kotov
Sergei Revin
Sergei Kononenko

Civilians, 1997
Oleg Skripotchka
Fyodor Yurchikin
Sergei Moshchenko

Pilots 1997
Dmitri Kondratyev
Yuri Lonchakov
Oleg Moshkin
Roman Romanenko
Alexander Skvortsov
Maxim Surayev
Konstantin Valkov
Sergei Volkov
Valeri Tokarev

Civilians, 1998
Yuri Baturin (presidential advisor)
Yuri Shargin
Mikhail Kornienko

Pilots, 2003
Anatoli Ivanishkin
Alexander Samokutyayev
Anton Shkaplerov
Yevgeni Tarelkin

Civilians, 2003
Oleg Artemyev
Andrei Borisenko
Mark Serov
Sergei Zhukhov

Doctor/biochemist, 2003
Sergei Ryazansky

Pilots, 2006
Alexander Misurkin
Oleg Novitsky
Alexei Ovchinin
Maksim Ponomarev
Sergei Ryzhikov

Civilians, 2006
Elena Serova
Nikolai Tikhonov

Some of these names were familiar, for they comprised the sons of cosmonauts: Sergei Volkov, son of Alexander; and Roman Romanenko, son of Yuri. Sergei Ryazansky was son of Mikhail, one of the designers of the 1946 Council of Designers. Later, two Kazakh pilots joined the 2003 group: Mukhtar Aimakhanov and Aidyn Aimbetov (marked Kz in Table 6.2). Within a few months, the new trainee cosmonauts were pictured flying jetplanes, doing weightless training and parachuting out of Mil helicopters.

By 1994, principally due to the cancelation of the Buran program, the cosmonaut squad had contracted to 34 members (17 pilots, 12 engineers and five doctors), the lowest number since the mid-1960s. Ten years later, by 2004, the squad had built up to 40 and stabilized at around this level, about half being pilots, half engineers. The

Table 6.2. Russian Cosmonaut Squad, 2007

Air Force and Space Force	Energiya	Others
Viktor Afanasayev	Alexander Kaleri	Sergei Moschenko
Yuri Baturin	Oleg Kononenko	Boris Morukov
Dmitri Kondratyev	Mikhail Kornienko	Sergei Ryazansky
Oleg Kotov	Konstantin Kozeyev	Sergei Zkukov
Yuri Lonchakov	Sergei Krikalev	Aidyn Aimbekov (Kz)
Yuri Malenchenko	Alexander Lazutkin	Muktar Aimakhanov (Kz)
Gennadiy Padalka	Sergei Revin	
Roman Romanenko	Oleg Skripochka	
Salizhan Sharipov	Sergei Treshev	
Alexander Skvortsov	Mikhail Tyurin	
Maxim Surayev	Pavel Vinogradov	
Valeri Tokarev	Fyodor Yurchikin	
Konstantin Valkov	Mark Serov	
Sergei Volkov	Andrei Borisenko	
Yuri Shargin	Oleg Artemyev	
Anatoli Ivanishkin	Elena Serova	
Alexander Samokutyayev	Nikolai Tikhonov	
Anton Shkaplerov		
Yevgeni Terelkin		
Aleksandr Misurkin		
Oleg Novitsky		
Alexei Ovchinin		
Maksim Ponomarev		
Sergei Ryzhikov		

Russian cosmonaut squad was the second largest in the world—far behind the United States with its large number of shuttle pilots and mission specialists (132), but well ahead of Europe (13), China (12) and Japan (8). The make-up of the cosmonaut squad can be seen in Table 6.2.

Of the two 1992 selections, Lazutkin and Vinogradov got early assignments, Lazutkin finding himself on Mir during the 1997 collision and Vinogradov on the subsequent repair mission. Mikhail Tyurin followed later. Nadezhda Kuzhelnaya never got to fly. One of the small number of women in the cosmonaut squad, she waited repeatedly in line for a mission but never got a final, definitive assignment and eventually left in frustration in 2004 to fly planes for Aeroflot. Although the Russians had been first to send women into space, their subsequent record on equal opportunities could hardly have been much worse. Only three Russian women have now flown in space, in contrast to the Americans, where over thirty have now flown. A woman, Eileen Collins, bravely led the return to flight of the space shuttle in 2005 when she commanded the *Discovery* shuttle. A woman candidate, Anna Zavyalova, actually passed all the tests for the 2003 civilian selection, only to be refused on the last hurdle, the State Commission, no satisfactory reason being given. The cosmonaut profession continued to be a far from equal playing field for women and female American astronauts in Star Town found Russian male attitudes to women to be quite outdated [13].

Being an unflown member of the cosmonaut squad during this period must have been a trying experience, granted the small number of flying opportunities likely to arise. Generally, the three-seat Soyuz would fly two Russians, with the third seat being taken by a visitor, but, with the financial situation still tight, two visitors might well take two of the three seats. The first seat, that of the commander, would invariably go to an experienced cosmonaut. Granted that there were only two Soyuz missions a year, with only one seat guaranteed each time for an experienced Russian, unflown cosmonauts were likely to have a long wait. Essentially, new cosmonauts found themselves in a situation where they were likely to be bumped by any foreigner, space tourist or foreign space agency able to afford to buy their seat. Soyuz TMA-5 was a case in point, with experienced Salizhan Sharipov taking the first seat, American Leroy Chiao the second and tourist Gregory Olsen the third. Only when Gregory Olsen was (temporarily) medically disqualified did the third seat go to a Russian, the lucky Yuri Shargin.

Cosmonauts launched, 2000–6 (Russian rockets)

6 Apr 2000	Soyuz TM-30	Sergei Zalotin
		Alexander Kaleri
31 Oct 2000	Soyuz TM-31	Yuri Gidzenko
		Sergei Krikalev
30 Apr 2001	Soyuz TM-32	Talgat Musabayev
		Yuri Baturin
21 Oct 2001	Soyuz TM-33	Viktor Afanasayev
		Sergei Kozeyev
25 Apr 2002	Soyuz TM-34	Yuri Gidzenko
28 Oct 2002	Soyuz TMA-1	Sergei Zalotin
		Yuri Lonchakov
26 Apr 2003	Soyuz TMA-2	Yuri Malenchenko
18 Oct 2003	Soyuz TMA-3	Alexander Kaleri
19 Apr 2004	Soyuz TMA-4	Gennady Padalka
14 Oct 2004	Soyuz TMA-5	Salizhan Sharipov
		Yuri Shargin
15 Apr 2005	Soyuz TMA-6	Sergei Krikalev
1 Oct 2005	Soyuz TMA-7	Valeri Tokarev
30 Mar 2006	Soyuz TMA-8	Pavel Vinogradov
18 Sep 2006	Soyuz TMA-9	Mikhail Tyurin

A small number of cosmonauts had the opportunity to fly on visiting shuttle missions to the International Space Station. These were as follows:

Cosmonauts on the shuttle (individual missions)

19 May 2000	*Atlantis* STS-101	Yuri Usachov
8 Sep 2000	*Atlantis* STS-106	Yuri Malenchenko
		Boris Morukov
19 Apr 2001	*Endeavour* STS-100	Yuri Lonchakov
7 Oct 2002	*Atlantis* STS-112	Fyodor Yurchikin

Cosmonauts on the shuttle (expedition crews up to ISS)

8 Mar 2001	*Discovery* STS-102	Yuri Usachov
10 Aug 2001	*Atlantis* STS-105	Vladimir Dezhurov
		Mikhail Tyurin
5 Dec 2001	*Endeavour* STS-108	Yuri Onufrienko
5 Jun 2002	*Endeavour* STS-111	Valeri Korzun
		Sergei Treschev
24 Nov 2002	*Endeavour* STS-113	Nikolai Budarin

Although the Russians had recruited medical doctors for space missions from as far back as 1964 (Dr. Boris Yegorov was first to fly), very few doctors had actually got missions (Oleg Atkov on Salyut 7, Valeri Poliakov on Mir). Boris Morukov was only the fourth and he flew on the American shuttle.

Sadly, the early years of the new century also saw the passing of a number of Russia's older, senior cosmonauts: Gherman Titov (2000); Vladimir Vasyutin (2002); Nikolai Rukhavishnikov (2002) and Oleg Makarov (2003), both of whom should have flown to the moon; Andrian Nikolayev (2004); and the much-liked Mir veteran Gennady Strekhalov (2004). Another widely regretted death on 21st May 2003 was that of Yaroslav Golovanov, a journalist who had done much to uncover the secrets of the Soviet space program and who had himself, at one stage, trained for a journalist-in-space mission.

Cosmonaut #2, Gherman Titov, was only 65 when he died in a domestic accident: carbon monoxide poisoning while in a sauna. He had been senior serving cosmonaut since the death of Yuri Gagarin in 1968 and was given a state funeral, with burial in the Novodevichy Cemetery. Vladimir Vasyutin died of cancer in Kharkov, Ukraine, aged just 50. Andrian Nikolayev, who after Titov had become the most senior cosmonaut, died of a sudden heart attack aged 74 while attending a sports event and had hitherto been in good health. The senior cosmonaut now is his colleague from the 1962 double mission, Pavel Popovich. Finally, chief designer Vasili Mishin died on 10th October 2001 at 84, much missed by journalists to whom he had willingly related the story of the Soviet side of the Moon race.

MISSION CONTROL KOROLEV: TsUP

The Apollo Soyuz Test Project was the first occasion in which Westerners penetrated the inner sanctuary of Star Town—indeed, the facilities built for American visitors then were later converted into the town's health center. Similarly, the project permitted the first American access to the flight control center in Kaliningrad. The existence of a major control facility in the area had not been acknowledged till then—indeed, maps of the area were not available until the late 1990s.

Kaliningrad was, before the revolution, a forest north of Moscow where wealthy Muscovites had country houses—there were about 50 in the area and Lenin came to live in one of them from January to March 1922 before his final decline. The center of the district was the government's forestry institute, located there in 1890. At that time it was called Podlipki and a railway station opened there in 1914. The railway station alone retained its name in the face of subsequent revolution and counter-revolution. Podlipki was renamed Kalininsky district in 1928 and then Kaliningrad in 1938 after the Soviet Union's first president Mikhail Kalinin (formal head of the Soviet Union from 1919 to 1946). Kaliningrad was renamed Korolev after the greatest of the great designers on 8th July 1996 by President Yeltsin, its third name that century.

Kaliningrad was eyed by the government as a center for the development of rocketry as far back as 1940 when rocket gliders were tested there. Arms factories were moved there in autumn 1941 when the Germans moved in on Leningrad, though they were in turn evacuated to the Urals when the Germans reached Moscow later that year. The factories were rebuilt there in 1946, the principal one being OKB-1, Korolev's design bureau. The A-4 Germans lived there briefly, before being moved to Seliger lake; the security-conscious nature of the area was emphasized by the building then of kilometers of brick walls. By 1960, there were eight design bureaus and rocket factories in Podlipki employing 200,000 people: a magnet for snooping American spyplanes—indeed, the brick walls marked out their contours nicely for them. Now, many of the main design bureaus may be found there, such as Energiya, Strela and Zvezda. Indeed, one of the ironies of the great rivalries between the design bureaus in the 1960s—which partly cost the USSR the moon race—was that some were in close physical proximity to each other.

Until 1973, mission control for Soviet space missions had been Yevpatoria in the Crimea, whose location and role had always been acknowledged. Korolev had been a frequent visitor there for the interplanetary missions. Yevpatoria had been supplemented by and linked to other tracking facilities in the 1960s, enabling cosmonauts to talk to the ground over Soviet territory: Dzhusaly, Kolpashevo, Tbilisi, Ulan Ude, Ussurisk and Petropavlovsk. They got a few minutes talking time as they flew over, what the Americans call a "communication pass" or "comm pass". But, Yevpatoria was a long way from the hub of Soviet space activity in Moscow.

So, mission control, or TsUP (*Tsentr Upravleniye Polyotami*), pronounced "Tsoop", was built in Kaliningrad for the Apollo Soyuz Test Project in 1975, though in the event it came online in September 1973 when it handled the Soyuz 12 mission. It is not unlike Western mission control centers, with six banks of 20 consoles per row, central wall display, a television link to the space station and maps of operational tracking stations. There are two very similar, large control rooms there: the main mission control, which traditionally handled the Mir space station; and Buran control, which masterminded its 1988 mission. During the 1990s the Buran control room was converted to the mission control center for the International Space Station, and this became the part of Korolev best known to Western viewers. The old, 1973 facility is now used for Soyuz launches, dockings and reentries.

TsUP

There are two smaller rooms: one to handle video linkups between the cosmonauts in orbit and their families; and another to control Progress and Soyuz missions when flying independently. TsUP controlled the International Space Station until February 2001, when it passed to Mission Control Houston, but there was never an official handover: it just happened over time. On three occasions, though, control passed back to Moscow: the 11th September attacks on the United States and twice when hurricanes Lili and Rita threatened Houston. Two thousand people work in TsUP, of whom 300 are mission controllers. Mission controllers work a pattern of 24 hr on, 72 hr off.

An American support room opened temporarily adjacent to the main control room during the Apollo Soyuz Test Project in 1975 and this was re-opened for the shuttle missions to Mir in the 1990s. With the ISS, this re-opened permanently as the "Houston Support Group" in 1998. Its role is to coordinate space station control between Korolev and Houston, especially before the day shift begins in Texas. Joint meetings of the controllers are held every Thursday morning. Their job is to reconcile the ISS workplans of the two partners, what the Americans call short-term plans and the Russians call cyclograms. Cyclograms used traditionally to be pencil-written on 2 m long sheets of paper, but were recently computerized. NASA operates what is called the Moscow Technical Liaison Office, with staff in TsUP and Energiya, which has developed standardized forms and procedures for resolving routine issues between the two sides (e.g., the cargo list for the next Progress freighter) as well as problem issues that may arise [14].

The European Space Agency opened, for the Eneide mission in April 2005, an ESA mini-mission control room, called a "support room", with its own control panels and booths which will, in the future, be the Moscow end of European-manned operations on the International Space Station once the European Columbus module arrives.

MILITARY MISSION CONTROL

A number of centers are responsible for the control of military missions. The main military tracking center is in Golitsyno. Like many such facilities, it went through many changes of nomenclature, starting as facility #413, then Golitsyno-2 and now Krasnoznamensk, its current name. Officially, of course, it did not exist during the Soviet period (it was never marked on maps) and formal information on Golitsyno was not published until 2000, even though the first buildings had gone up as long ago as 1957 and as many as 30,000 people live there now, of whom 2,000 work directly in the control center.

Golitsyno is 41 km west of Moscow on the highway to Minsk. It was originally the coordinating station for the national tracking system set up to follow Sputnik. The center is full of small consoles, with none of the big screens to be found at Korolev. Golitsyno can control up to 120 satellites at a time and can handle huge quantities of data. In August 2000 the center was equipped with mobile command posts, meaning that its functions could be dispersed in the event of conflict. Although not formally involved in the civilian space program, it tracked the Zarya space station during its first few minutes in orbit and may have saved it from some untimely problems. Following the death of cosmonaut Gherman Titov, the center was named after him. President Putin gave French president Jacques Chirac a tour in April 2004 [15].

The Oko early-warning system is controlled from another control center in Kurilovo, 100 km southwest of Moscow on the road to Kaluga, its military codename being Serpukhov 15. This was out of action from May to September 2001 as a result of a bad fire, with serious consequences for the early-warning system. The naval observation system of US P satellites is run from another center, Noginsk.

TRACKING AND CONTROL

The scale and scope of the Soviet space program required a national and international network for the tracking, control and recovery of spacecraft. The original Soviet tracking system dated to September 1956, when nine measurement centers or instrument points (in Russian, IPs) were constructed in the downrange path of the R-7 missile, so as to follow its direction and ascertain whether it reached its target accurately [16]. To follow the first artificial satellite, Sputnik, construction of another four instrument points began the following year: these were called scientific instrument points (NIPs). These were later expanded to twenty-one, as follows:

 1   Baikonour
 2   Makat, Guriev
 3   Sari Shagan, Lake Balkash
 4   Yeniseiesk, Krasnoyarsk
 5   Ishkup
 6   Elizovo, Kamchatka
 7   Barnaul, Kluchi, Siberia
 8   Bolshevo, Moscow
 9   Krasnoye Selo, St. Petersburg
10   Simferopol, Crimea
11   Tbilisi
12   Kolpashevo, Novosibirsk
13   Ulan Ude, Lake Baikal
14   Shelkovo, Moscow
15   Ussurisk, Vladivostok
16   Yevpatoria, Crimea
17   Yakutsk, Siberia
18   Vorkuta, near Plesetsk
19   Dunavetsu, Moldova
20   Solnechny, Komsomolsk, Amur
21   Maidanek, Uzbekistan

The original Soviet space surveillance system, called SKKP (System for Monitoring Space) was set up in 1962, with radio, radar and optical devices and, later, lasers. Radar-tracking stations were set up in Irkutsk, Murmansk, Pechora, Sevastopol, Uzhgorod, Balkash, Mingechaiur and Riga, but only the first three—still being on Russian territory—are now available.

This was superceded by the Okno system ("Okno" is the Russian word for "window"), which began in 1969 in a R120m effort to spot American spy satellites as they tracked over Soviet territory as well as to keep track of Soviet satellites as far out as geostationery orbit 36,000 km high. The contract went to the KMZ company in Krasnogorsk, maker of the Zenit cameras on the early photo-reconnaissance satellites (also called Zenit) and a range of military optics from night vision glasses for soldiers to fire control systems on tanks. Chief designer at KMZ was 40-year-old Vladimir Chernov who recruited a young team of new graduates for the task, led by Valeri Kolinko.

Okno was built quite quickly and set up for testing at Zvenigorod Observatory near Moscow in later 1969. That autumn was exceptionally cloudy and, after months of patient waiting and growing frustration, it was decided to move the test to Byurokan Astrophysical Laboratory in Soviet Armenia, far away but cloudless by night and offering a better view of the skies from higher altitudes. The Leningrad Design Bureau of Special Machine Building was responsible for the base of the system and used—instead of the traditional ball bearings—a 100-micron thin layer of oil under intense pressure (70 atmospheres), sufficient to hold the 15-tonne weight of each telescope. The project even involved KGB agents obtaining high-sensitivity television equipment from abroad.

Large tracking dish

Okno was successfully evaluated over 1969–72 and approval for an operational system was given in 1974. The top of Sanglok Mountain—2,200 m high, near Nurek in Tajikistan and 80 km southeast of the Tajik capital Dushambe—was leveled, the telescope installed and trees planted round about. Construction did not begin until 1979 and took ten years. For the Soviet Union, Sanglok was a perfect site, for most American satellites tracking across the Soviet Union came within its range at the southern extreme of the USSR. Commissioning was in progress in 1991 when Tajikistan declared its independence from the Soviet Union. Civil war broke out three years later and the electronic engineers manning the project had to bring in combat experts to train them to use weapons to defend their precious facility from attack. Another three years later and the Russian Federation abandoned the expensive project. KMZ had almost collapsed, water was pouring in through the factory roof and Chernov had retired. His replacement, Valeri Kolinko, did not get the message that the project was over and in 1999 persuaded the Russian government to re-open the facility, even though Western computer systems had to be used at first. It came back on stream in March 2004. Valeri Kolinko won a state prize for his perseverence.

Okno now uses ten telescopes sheltered in 25 m wide domes to track and catalog all satellites passing overhead as well as those in geostationery orbit and out to 40,000 km. Computer systems are able to eliminate all stars on the image, leaving only satellites. Okno can swivel to a full range of angles in the sky, operating at short, medium and long range. Within weeks of opening, it was able to warn of an old American satellite approaching close to the International Space Station. The staff

Tracking dishes in snow

there work only at night, for the mirrors would be destroyed by exposure to daylight. At an agreement in Dushambe in 2004, the Tajik government ceded the facility to the Russian government as Russian territory in exchange for the waving of €200m in outstanding debt. The 360 clear days a year there give the best possible viewing. Although conditions in Tajikistan have calmed, Afghanistan is only 60 km away and the Okno facility includes what is called an "anti-sabotage squad" of elite soldiers for protection [17].

Like the Americans, the Russians now keep a full catalog of objects circling the Earth. The number of objects tracked has risen from 250 a year in 1969 to 1,000 a year in 1975 and 7,500 in 1994. Other optical tracking equipment is operated in Zvenigorod, Moscow and Maidenek, Uzbekistan.

Before the opening of Kaliningrad mission control, the main control center was in Yevpatoria, western Crimea, chosen in 1957 by Korolev himself, offering a southerly latitude. The first tracking stations were built there in 1958 in time for the first Moon launches that autumn, but the main complex came on line in September 1960 for the first launches to Mars. The eight original 16-m tracking antennas were built on converted battleship turrets. It was called the TsDUC, or Center for Long-Range Space Communications. TsDUC actually comprised two centers: a western one (Yevpatoria, which received its first Western visitors in 1963) and an eastern center in Ussurisk, near Vladivostok, called the Pluton system. They were joined later in the 1960s by a network of 32-m Saturn dishes (Yevpatoria, Baikonour, Sari Shagan, Shelkovo, Yeneseiesk) and in the late 1970s by 70-m Kvant dishes in Bear's Lake (Moscow), Yevpatoria, Kalyashin on the Volga and Ussurisk. During the period of the USSR, all lunar and interplanetary missions were controlled from Yevpatoria, supplemented by the Saturn and Kvant systems. After Ukrainian independence, the

Yevpatoria system briefly withdrew from the network, but subsequently returned. This system had little use in the period after the fall of the Soviet Union, but is expected to be pressed back into service with the resumption of lunar (Luna Glob) and Mars missions (Phobos Grunt).

An important feature of the Soviet tracking system was the comships, or communications ships, used. Comships were used to track lunar and deep-space missions and maintain contact with cosmonauts when their orbits took them far from the USSR. Large tracking ships were constructed for the moon effort—the first being the *Cosmonaut Vladimir Komarov* (17,500 tonnes), followed by the *Cosmonaut Yuri Gagarin* (45,000 tonnes), which became the flagship and then the *Academician Sergei Korolev* (21,250 tonnes). Another large ship, the *Academician Nikolai Pilyugin*, was laid down in Leningrad in April 1988. They were impressive, streamlined, white ships and with their giant aerials and huge telescope-like domes they looked futuristic. A series of smaller tracking comships was commissioned in 1974—the *Cosmonaut Pavel Belyayev* (1978), the *Cosmonaut Georgi Dobrovolski* (1978), the *Cosmonaut Viktor Patsayev* and the *Cosmonaut Vladislav Volkov* (1977). They were accompanied by a fleet of smaller, converted comships—*Borovichi*, *Kegostrov*, *Morzhovets* and *Nevel*. The Soviet navy also commissioned its own comships, presumably for use in association with its naval reconnaissance satellites. These were the *Marshal Mistrofan Nedelin*, assigned to the Pacific fleet and the *Marshal Krylov*.

In 1992, as the economic crisis began to bite in Russia, all the tracking ships were recalled, even though this meant that the then-orbiting Mir cosmonauts were now out of touch with ground control for up to nine hours at a time. The *Borovichi*, *Kegostrov*, *Morzhovets* and *Nevel* were sold and the *Cosmonaut Vladimir Komarov* became, briefly, an environmental monitoring ship in the Gulf of Finland. In 1994 the *Cosmonaut Yuri Gagarin* and *Academician Sergei Korolev*, after lying idle in Odessa for some time, came under the control of the Ukrainian Space Forces who at once tried to sell them—but no buyers appeared. Eventually, in 1996 the *Cosmonaut Yuri*

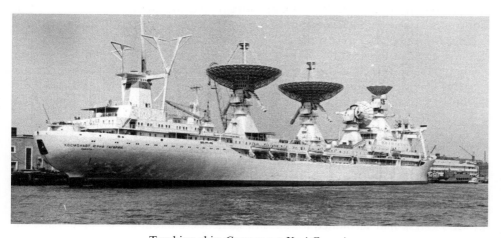

Tracking ship *Cosmonaut Yuri Gagarin*

Gagarin, the *Academician Sergei Korolev* and the *Cosmonaut Vladimir Komarov* were scrapped at a price of $170 a tonne. Russia could not afford to put a tracking ship on station for even a week for the Mars 8 launch in 1996. Following a decision by the federal government's property management agency, the *Cosmonaut Georgi Dobrovolski* was sold in St. Petersburg for scrap for R24m to an Antilles Islands company in December 2005. Several people protested and argued that the ship could be quickly made operational once again.

At this stage only one ship remained: the *Cosmonaut Viktor Patsayev*. It was converted into a floating museum in Kaliningrad, formerly the German Baltic city of Königsberg. During Hurricane Rita in the Gulf of Mexico, mission control in Houston had to go off-line and the *Cosmonaut Viktor Patsayev* quickly whirred back into action and briefly served as a backup mission control.

With the end of the tracking fleet and the demise of the short-lived Luch network of space-based relays (Chapter 3), mission control became reliant on ground stations alone to communicate with its cosmonauts. However, not all are in the Russian Federation, and consequently the tracking network was in danger of shrinking even further. As a result, Russia's ability to communicate with its orbiting spacemen and women was even more restricted than it had been when Yuri Gagarin first circled the Earth.

Attempts to develop a land-based system abroad emerged in the course of a 2006 agreement signed between Roscosmos and the government of South Africa. The enterprising Chinese already had a ground station in neighboring Namibia to follow their manned Shenzhou spacecraft during the critical moment of reentry over southwest Africa. A possible candidate for the ground station was a satellite control center secretly built by the South African government during the apartheid period in the 1980s, but never actually used. It has a great location, less than two hours from Cape Town, in pine plantations and apple orchards [18].

For Meteor weather satellites, three tracking stations were set up: Obninsk (near Kaluga); Novosibirsk and Kharbarovsk, which transferred their data to the World Meteorological Center in Moscow for redistribution throughout the country. Later, a fourth station was built at Dolgoprodny, near Moscow, where the main data storage center was located. The system was run, from 1997, by the NITsPlaneta, the Scientific Research Center of Space Meteorology, which takes in both Russian and foreign data through these stations and through over 60 automatic stations scattered all over the country. The actual control of the weather satellites was run from Golitsyno in the 1990s, but transferred to mission control in Korolev (TsUP) in time for Meteor 3M1 [19].

COSMODROMES AND GROUND FACILITIES: CONCLUSIONS

From 2000 to 2006, Plesetsk had 39 launches, Baikonour had 137, *Odyssey* 20, Svobodny three and the Barents Sea and Dombarovska one. During the Soviet period, Plesetsk had accounted for 55% of launches, Baikonour 40% and Kapustin Yar 5%. Now, Baikonour is the clear leader with 66%, Plesetsk 20%, *Odyssey* 10%

and others 4%. This difference may be largely attributable to the fall in the launch rate of military satellites, which were concentrated on Plesetsk and the growth of commercial satellite launches, almost all of which took place at Baikonour and which had scarcely been a feature of the Soviet program at all.

Like the rest of Russia's space program, the cosmodromes went through difficulty and hardship in the 1990s. Overall, the physical conditions of Baikonour and Plesetsk declined. However, parts of Baikonour were modernized and are now a busy, international spaceport that will service commercial operators and the International Space Station for at least another ten years. Although the relationship with Kazakhstan over Baikonour went through a turbulent period, this settled down. Despite most military launches moving to Plesetsk, Baikonour remains active as the core of Russian launcher activity.

Russia was able to maintain its principal ground facilities in the 1990s—Star Town, Korolev and mission control. The mission control center was adapted to serve the new International Space Station. The one feature of its ground infrastructure which suffered most was the tracking system, which almost collapsed. Russia lost the automatic use of tracking stations outside the Russian Federation. Its inability to maintain the Luch network was a handicap in the operation of the ISS. Most importantly, Russia lost its entire, once proud, tracking fleet and its many fine ships. They were recalled early in the 1990s and none ever set sail again. Meanwhile, the Chinese had four splendid, new, big tracking ships in their fleet, playing an important part in China's new manned space program—a contrast of changing fortunes.

REFERENCES

[1] William P. Barry: *The missile design bureaux and Soviet manned space policy, 1953–1970*. Doctoral thesis, Merton College, University of Oxford, 1996.

[2] Speth, Roland S.: The Baikonour cosmodrome. *Spaceflight*, vol. 43, April 2001.

[3] Harpole, Tom: White elephant. *Air & Space*, December 2002/January 2003.

[4] George C. Larson: Leroy's launch. *Air & Space*, June/July 2005; Sotham, John: Baikonour. *Air & Space*, February/March 2001.

[5] De Angelis, Laurent: Soyuz in the jungle, from Brian Harvey (ed.): *2007 Space exploration annual*, Springer/Praxis, 2006.

[6] Strom, Steven R.: *International launch site guide*. The Aerospace Corporation with the American Institute of Aeronautics and Astronautics, El Segundo, California and Reston, Virginia, 2005.

[7] Mellow, Craig: Aiming for Arkalyk. *Air & Space*, August/September 1998.

[8] Lardier, Christian: La panne du Soyouz TMA-1 analysée. *Air & Cosmos*, #1893, 6 juin 2003.

[9] Da Costa, Neil: A private trip into space: Gregory Olsen—the third "space flight participant". *Spaceflight*, vol. 48, no. 2, February 2006.

[10] Grigoriev, Anatoli; Bogomolov, Vaery; Goncharov, Igor; Alferova, Irina; Katuntsev, Vladimir and Osipov, Yuri: *Main results of medical support to the crews of the International Space Station*. Paper presented to International Astronautical Federation, Valencia, Spain, 2006.

[11] Phillip S. Clark: Final equator crossings and landing sites of CIS satellites. *Journal of the British Interplanetary Society*, vol. 55, no. 1–2, January–February 2002.

[12] Rex D. Hall, David J. Shayler and Bert Vis: *Russia's cosmonauts—inside the Yuri Gagarin Training Center*. Springer/Praxis, 2005.

[13] Oberg, Jim: *Does Mars need women? Russians say no*. MSNBC, 11th February 2005.

[14] Covault, Craig: Station bridge. *Aviation Week & Space Technology*, 26th June 2006.

[15] Lardier, Christian: La Russie ouvre Golitsyno 2. *Air & Cosmos*, #1931, 9 avril 2004.

[16] Zak, Anatoli: Ground control stations (KIK), *www.russianspaceweb.com/kik*

[17] History of the Okno space monitoring facility. Russian television, Channel 1, broadcast 31st October 2006.

[18] Keith Gottshalk: South Africa—satellite command and control center. Posting on Friends and Partners in Space, 29th October 2006.

[19] Hendrickx, Bart: A history of Soviet/Russian meteorological satellites. *Journal of the British Interplanetary Society*, Space Chronicle series, vol. 57, no. 1, 2004.

7

The design bureaus

The organizational core of the Russian space program is the design bureau. This could be classified by design bureau (KB) or experimental design bureau (*Opytnoye Konstruktorskoye Buro,* OKB). The design bureau was the middle element in a three-part chain, a system developed in Stalin's time [1]. First, concepts were tested in a scientific research institute (NII) (*Nauk Issledovatl Institut*). Once deemed possible or desirable, hardware was designed, built and tested by an OKB or KB. Once perfected, it was put into production in the third part, the factory. The operation of the system was actually more complex than this, because some design institutes grew up with factories alongside and were closely associated with one another. Furthermore, a design product of an OKB could be sent for production in a factory affiliated to a rival design bureau. A complex set of relationships and rivalries thus built up over the years. Their work in the new century is now reviewed. Nowadays, many of these organizations are called NPOs (scientific and production associations), companies or corporations, but the term "bureau" is still widely used.

During the Soviet period, the great leaders of the program built up large design offices, factories and plants. They bid for contracts, put forward their own projects and embarked on rival enterprises. Designers (*konstruktor*) such as Sergei Korolev, Vladimir Chelomei, Mikhail Yangel, Semion Kosberg, Valentin Glushko and Dmitri Kozlov exerted considerable influence over Soviet planning and politics for a long period, their gigantic role unparalleled by industrial leaders in the United States, most of whom remained relatively anonymous. The design bureaus exercised power in their own right—indeed, it was the inability of the Soviet political machine to control them and their rivalries that was a major contributor to the Soviet Union losing the Moon race.

The Soviet design bureaus, the organizational bedrock of the Soviet space program, survived the transition to the Russian space program. The largest, Energiya, experienced the most financial difficulties, but remained undisputedly preeminent. Effectively, it owned the manned space program, the cosmonauts, mission control

Early group of Soviet space designers

and the Russian part of the International Space Station project. Its rival of old, the Chelomei design bureau, was associated with the Khrunichev company, which, continuing in state hands, became a profitable international space corporation selling Proton rockets on the world market and marketing new models, like Rockot. Accelerated by the break-up with the Ukraine, the NPO Yuzhnoye became a less important element, while still supplying key rockets and satellites, especially for the military program. Other design bureaus, like TsSKB, Lavochkin and NPO PM, managed to adapt by developing their existing products and diversifying into new areas.

The Russian space program, the least commercialized in the world during the Soviet period, quickly became the most commercialized. Although the process of transition had begun in the final two or three years of the rule of Mikhail Gorbachev, commercialization proceeded with a vengeance from early 1992 onward. Rather than allow the program to sink without trace in financial collapse and bankruptcy, the leadership of the Russian space program and the design bureaus quickly re-orientated themselves around the new economic realities. While some aspects of commercialization were clumsy, most Russian space companies managed to arrange joint ventures or other forms of partnership with Western companies. The outcomes were uneven, with modest results for some companies and substantial successes for others (e.g., Khrunichev, Energomash). Several design bureaus and companies, like Energiya, built up significant export earnings. Russia developed what the Soviet Union never had: a commercial space program in the global capitalist economy. Table 7.1 lists the main design bureaus over 2000–6 and their areas of expertise.

ENERGIYA—PREMIER DESIGN BUREAU

The premier design bureau from the Soviet period onward was Design Bureau #1, the original OKB-1 in Podlipki, directed by Sergei Korolev, who designed the R-7, Vostok, Voskhod, Soyuz, the N-1, the Zenit spy satellites and the first generation of

Table 7.1. Main design bureaus and agencies in the Russian space program

Company name	Location	Area of expertise, product
Energiya (Korolev)	Korolev, Moscow	R-7, Soyuz, Progress, ISS, Kliper, Parom, block D, Yamal
NPO Energomash (Glushko)	Moscow	Rocket engines: RD 107, 108 series RD-253 series, 170 series, 190 series RD-214 series
Energomash, Volga branch	Samara	Onega, Yamal, Aurora
Khrunichev (Chelomei)	Moscow	Proton, FGB, Briz KM and M, Angara, Rockot, Monitor
TsSKB Progress (Kozlov)	Samara	Yantar, Resurs DK, Foton, Orlets
NPO Lavochkin (Babakin)	Moscow	Fregat, Meridian, Araks, Oko, Prognoz, Kupon, Spektr, Phobos Grunt, IRDT, Luna Glob
Makeev	Miass, Chelyabinsk	Volga, Shtil rockets
MIT Kompleks	Moscow	START
VNIIEM	Moscow	Meteor
NPO PM (Reshetnev)	Krasnoyarsk	GLONASS, Luch, Ekspress, Ekran, Molniya, Raduga, Strela, Gonetz, Potok
NPO Polyot	Omsk	Cosmos 3M
NPO Yuzhnoye/Pivdennie (Yangel)	Dnepropetrovsk	Zenit, Tsyklon, Dnepr, Sich, Tselina, Koronas
KB Arsenal	St. Petersburg	US P
OKB Fakel	Korolev, Moscow	Electric engines
KBKhA (Kosberg)	Voronezh	Rocket engines, RD-0124
KB Khimmash (Isayev)	Korolev	Rocket engines, KVD-1
IZMIRAN	Moscow	KOMPASS
Keldysh Centre	Moscow	Rocket engines
KBTM	Moscow	Cosmodromes
Pilyugin Centre	Moscow	Rocket and satellite guidance systems
NPO Zvezda	Moscow	Spacesuits
Institute of Space Device Engineering	Moscow	Radio and telemetry systems

Sergei Korolev

lunar and interplanetary probes. When he died, OKB-1 became NPO Energiya under his successors Vasili Mishin (1966–74), Valentin Glushko (1974–88) and then Yuri Semeonov (1989–2004). In 1994 it was renamed *RKK Energiya imemi Sergei Korolev* (Rocket Cosmic Corporation Energiya, dedicated to the memory of Sergei Korolev). The company was part-privatized in 1994, the government keeping 51% of the holding, but a further 13% was sold off three years later to raise funds to launch the Zvezda service module. It is still by far the largest design bureau and employs between 22,000 and 30,000 people.

Energiya, located in the town of Korolev, remains predominant among the Russian design bureaus, is responsible for all manned-related operations and owned the Mir space station. In 1996 it published a coffee table book history of its first 50 years and its key role in Soviet and Russian space exploration was readily apparent. The bureau is the lead organization for the development of the Soyuz, Progress and space stations. Most of those working in mission control in Korolev belong to the Energiya bureau, which is likely to maintain its high visibility for some time. It was the Energiya engineering staff who sorted out the problems on Mir in 1997 and its managers who worked hard and largely successfully to attract foreign investment for the Mir station throughout the decade. Almost half of Russia's cosmonauts are recruited from Energiya engineers.

Nikolai Sevastianov

Energiya is the lead agency for the Russian end of the space station project. The company bore the brunt of the financial shortages which crippled the Russian space program in the late 1990s. Its managers struggled endlessly with deficits by cajoling money from politicians, begging advances from NASA and by privatizing ever more of the company's stock. Against the odds, they managed to launch the Zvezda module to the International Space Station in July 2000. Following a boardroom battle, Nikolai Sevastianov became the fifth director of Energiya in 2004, favored both by shareholders and by President Putin, ousting Yuri Semeonov.

Energiya actually turned a profit in 2005, the first year ever, just €9m, but sufficient for Sevastianov to recommend 20% go to the shareholders. Not a lot, but better than the old days, when the chief designer's slide rule had to be sold at public auction to prevent repossession.

CHELOMEI'S BUREAU AND DERIVATIVES

The main rival to Korolev was OKB-52, located at Reutov, Moscow in August 1955, directed by Vladimir Chelomei, who masterminded the building of the Almaz space station, the Proton rocket and a range of spaceplanes and military projects. During the big re-organization of the space program in 1974, it was renamed NPO Mashinostroeniye. The bureau suffered severely from the funding cuts of the 1990s and by 1997 had only 4,500 staff.

Responsibility for the Chelomei bureau's most lasting design, the Proton rocket, passed to the biggest rocket factory in Moscow, and the one best known in the West,

Vladimir Chelomei

the neighbouring MV Khrunichev State Research and Production Space Center. Originally an automobile plant, then a production line for German Junkers monoplanes, in the war a producer of Red Air Force planes, it became the Myasishchev Design Bureau and then the Salyut Design Bureau. From 1960, it received contracts for the production of rockets and spacecraft designed by Vladimir Chelomei and, most important of all, the Proton rocket, from 1965.

Khrunichev was one of the first companies to enter an arrangement with Western companies, when it signed a deal with Lockheed Martin in 1993 establishing the International Launch Services (ILS) Joint Venture. As a result, Khrunichev was soon able to attract significant foreign contracts for the launching of commercial and communications satellites with the Proton rocket. Capital investment from Lockheed enabled Khrunichev to modernize its facilities, especially those for launch preparation in Baikonour. Ironically, it was the privatized Energiya company which experienced the greater economic difficulties in the 1990s, while the Khrunichev company, which continued to be a state enterprise, prospered from steady funding from its foreign ventures. The FGB module for the International Space Station was built there—on schedule—and the first elements of the new Angara rocket began to appear there in 2000. Khrunichev also developed the new Briz upper stage and the Rockot smaller launcher. Khrunichev made the transition from factory to design bureau by designing and building the new, small Earth resources satellite, Monitor.

NPO LAVOCHKIN

As OKB-1 became overloaded with work in the mid-1960s and the urgency to beat America became overwhelming, many elements were hived off to new, distant or different design bureaus. The unmanned lunar program was first to be devolved—going to what had been the old Lavochkin Aircraft Design Bureau, OKB-301 in Khimki, which dated to 1937 but which had since become part of Chelomei's OKB-52. The bureau was reconfigured in 1965 under the guidance of Georgi Babakin, who

Georgi Babakin

was chief designer there from 1965 to 1971, succeeded by Sergei Kryukov (1971–78) and then Vyacheslav Kovtunenko (1978–95). Lavochkin had some spectacular successes, like the Luna moonscoopers, the Lunokhod moon rovers and the Venera landers on Venus. With the end of the lunar program and the virtual termination of planetary flights, one might have expected the Lavochkin bureau to sink out of sight, but it managed to hold on. The new upper stages Ikar and Fregat were built there, giving Lavochkin much new business in the late 1990s. It also obtained the contract for a number of military programs such as the Oko early-warning satellite, Prognoz and Araks, as well as the civilian program Kupon.

Stanislav Kulikov was the first Russian period appointment (1995), but he had the misfortune to arrive in post just before the Mars 96 mission. Subsequently a number of Lavochkin-built satellites suffered failures: the Cosmos 2344 and 2392, Araks optical reconnaissance satellite, the Kupon banking communications satellite, two Oko early-warning satellites and then a reentry demonstration test. Moreover, the head of the Russian Space Agency, Yuri Koptev, himself came from the Lavochkin bureau and he kept a paternal eye on its progress. He was angry with the run of failures there and in August 2003 dismissed Kulikov. Lavochkin staff, for their part, criticized Koptev for failing to deliver agreed budgets to the bureau, with the result that future missions had to be delayed or canceled. His position as chief designer was taken temporarily by Konstantin Pichkhadze and later by Georgi Polishuk.

One of Lavochkin's most innovative developments was that of an inflatable reentry demonstrator, called the IRDT (inflatable reentry and descent technology) and funded as an experimental project by the European Space Agency and the European aerospace and defence company EADS. American engineers had experimented with the idea of lightweight, rubbery structures and inflatables as a means of getting cargoes, experiments and even astronauts back from space, as far back as in the 1960s [2]. The idea of powering through reentry in a rubber raft was counterintuitive, but it was a good system for dissipating heat. Similar ideas had been developed in Russia and in 1997 the Babakin Research Center within Lavochkin approached the European Space Agency about the prospects for developing the idea, ESA finding €2m for such an experiment. The IRDT test was undertaken with a view to looking at whether a system could be developed to bring payloads down from the space station, enabling as much as 260 kg to be returned at a time and at a much more economic price than using the shuttle (Soyuz's return cargo capacity is only 50 kg). Inflatable cones have the great advantage that they can be packed into a small space, a ball of less than 80 cm, with a favorable weight-to-size ratio for the return cargo.

For the first IRDT test, there were actually two inflatable cones: the main one 14 m across and carrying the 1,800-kg Fregat motor; the second one was smaller, 4 m and within it a 110-kg payload, wired to measure the effects of reentry on the system. This carried different types of rock—basalt, dolomite and artificial cement/carbonate—to see how meteorites reacted to the heat of reentry. Both cones were made of silicone-based shields and stiff ablative material.

IRDT was first tested on the first flight of the Soyuz Fregat in February 2000. After five orbits, as they came in through reentry, each part deployed a rubber inflatable heat shield at 150 km designed both to protect the cargo during reentry and to cushion its impact on the ground. The effect of inflating the cones was to reduce the speed of reentry from 5,500 m/sec to 200 m/sec. At 30 km, the cones fully inflated in order to further reduce the speed of descent, now in the atmosphere. Final touchdown was to be cushioned by an inert gas-filled bag. Russian defence radars followed the reentry of the cargoes to a landing site near Orenburg.

The results were mixed. The larger cargo, the Fregat motor itself, was never found, although the telemetry suggested it did make it through reentry. Lavochkin blamed the loss on theft, alleging that someone resident in the steppe found the engine and melted it for scrap. The air bag did not deploy for the second, smaller, instrumented inflatable, which hit the ground at some speed. It was quickly found in the snow near Orenburg and, though badly damaged by the impact, thermometers suggested that the highest temperature it experienced was 25°C.

IRDT 2, weighing over 200 kg and with a 3.6 m diameter cone, was launched on 12th July 2002 from the submarine *Ryazan* under the Barents Sea using a Volna missile. It was sent high on a curving, high, 30-min ballistic trajectory. At its apex, the 80 cm diameter demonstrator was to detach, turn around into the flight path, inflate with nitrogen into a double set of shields and then form a protective shock wave as it punched through its reentry corridor. Helicopters fanned out over Kamchatka to find the IRDT, but there was no sign of its beacon and the search was abandoned at the end of the month. It is still not known if the payload overshot the target area, burned up or had deployed incorrectly or too early, or whether the failure took place much earlier, during the launching. The third IRDT, called 2R ("R" for replacement for number 2) in October 2005, likewise disappeared and it is not known if it landed, overshot or undershot, though the submariners claim both were launched properly.

Lavochkin also developed the solar sail mission. The mission has sometimes been termed Cosmos-1, but this designation is avoided here for risk of confusion with the small scientific satellite Cosmos 1 launched by Russia in March 1962. The sail was intended to be an operational test of solar sailing, an exciting, innovative mission. The first was a sub-orbital test, the second an orbital test, but both failed (see Volna, Shtil and relatives in Chapter 5, pp. 185–7).

Test missions, 2000–6: Lavochkin solar sail and IRDT

9 Feb 2000	IRDT 1	Baikonour	Soyuz Fregat
19 Jul 2001	Solar sail sub-orbit	*Borisoglebsk*	Volna
12 Jul 2002	IRDT 2	*Ryazan*	Volna
23 Jun 2005	Solar sail orbit	*Borisogblebsk*	Volna
9 Oct 2005	IRDT 2R	*Borisoglebsk*	Volna

Launches 2001 and after from the Barents Sea

NPO YUZHNOYE: MISSILE LINES "LIKE SAUSAGES"

Traditionally, the third largest design bureau in the Soviet system, after those of Korolev and Chelomei, was the NPO Yuzhnoye in Dnepropetrovsk in the Ukraine. To be precise, it comprises a connected design bureau (NPO Yuzhnoye) and factory (Yuzhmash), but they are discussed together here. Originally a car and tractor factory, a government resolution ordered the conversion of the plant to a rocket factory on 10th May 1951 under V.S. Budnik. A small design office was attached, originally as a branch of OKB-1 but acquiring its own identity on 10th April 1954 as OKB-586 under the leadership of Mikhail Yangel. To do so, the Soviet leadership posted 25 designers taken from both Korolev's and Glushko's design bureaus, to the strong protestations of both. After Sputnik, the Soviet leadership, especially Defence Minister Ustinov, saw no reason Korolev should have a monopoly of rocket and satellite building, so a party and government resolution of December 1959 charged Mikhail Yangel of the OKB-586 design bureau in Dnepropetrovsk to build a new generation of rockets and, in August 1960, a first run of ten small experimental satellites.

Mikhail Yangel

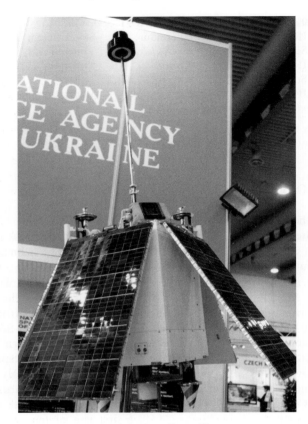

Ukrainian small satellite

OKB-586 built the R-12 (small Cosmos), R-14 (Cosmos 3), R-36 (Tsyklon) and Zenit rockets, though some construction was subsequently contracted out to other plants in Russia (e.g., the Cosmos 3M went to OKB Polyot in Omsk).

Mikhail Yangel developed a standard, mass-produced satellite design called the DS (*Dnepropetrovsky Sputnik*). The first was supposed to enter orbit in October 1961 to mark the 22nd Party Congress in Moscow, but the R-12 let them down and success was not achieved until Cosmos 1 in March 1962. No fewer than 125 DS sputniks were launched in their various sub-versions (DS U1, UJ2 and U3). In short order, Yangel began to develop a broad range of satellite programs, but in a big reorganization in the 1960s sea-based missiles were tranferred to the Makeev Design Bureau, weather satellites were assigned to the Iosifian Design Bureau and communication satellites to the NPO PM Reshestnev Design Bureau in Krasnoyarsk, to the fury of Yangel's two deputies, Budnik and Kovtunenko. Despite this, there was still plenty of satellite work for OKB-586, for the DS series was joined by a broad range of military satellites: RORSATs, EORSATs, the Tselina electronic intelligence satellite, Okean and Sich [3].

After Yangel's death in 1971, the OKB was led by Vladimir Utkin. In 1966 it was renamed NPO Yuzhnoye (the Russian word for "southern") and after Ukrainian independence Pivdennie (the Ukrainian word for "southern"), with the factory named Yuzhmash and Pivdenmash, respectively. Although the total number of programs handled by Yuzhnoye was not high, the production runs were often very, very long. At one stage about 50,000 probably worked in the space division, but that number is probably closer to 6,000 now. At one stage Yuzhnoye was running six rocket production lines simultaneously, so much so that Khrushchev once boasted to the West that "we produce rockets in a line like the way you make sausages."

After independence, Yuzhnoye continued to work with the Russian design bureaus and contractors with whom it had developed interlocking relationships for so many years, while at the same time attempting to promote an independent Ukrainian space program. Overall, though, its role in the Russian space program is diminishing, with the Tsyklon rocket near the end of its career. One of its most prominent activities at present is the manufacture of the Zenit 3SL rocket used for the Sea Launch program: about six come off the production line there every year, which will increase as orders for Land Launch come in. The Russian Federation will have much less need of the Zenit 2 once the Angara rocket goes into operation.

The precise role which the Ukraine was expected to play in the Russian space program has never been entirely clear. Ukraine's most important role was as the supplier of Zenit, Dnepr and Tsyklon rockets, flanked by a level of programmatic cooperation (e.g., Koronas program, Sich M) and infrastructural support (e.g., Yevpatoria tracking station, participation in GLONASS). At one level, Russia moved to ensure its independence from Ukraine-based systems, such as the expensive Kurs rendezvous system and the Zenit rocket: Russia began to develop its own system, Kurs M and its new range of Angara rockets. At the other level, President Putin made an early, February 2001 visit to President Kuchma of the Ukraine, the visit centered on the Yuzhnoye Design Bureau, where Leonid Kuchma had been factory director. The National Space Agency of the Ukraine marked the visit by an appeal for cooperation and joint programming between their two national space agencies. An intergovernmental agreement to govern the relationship between the two was agreed during their meeting and sent to the Ukrainian parliament for approval that summer.

With the orange revolution in Kiev at the end of 2004, the Ukraine began to consider a reorientation of its space program [4]. NPO Yuzhnoye opened an office in Brussels the following March and was soon appealing for the Ukraine to be permitted to join the European Space Agency. NPO Yuzhnoye Chief Designer Stanislav Konioukov soon met with ESA Director General Jean-Jacques Dordain and in no time there were three working groups between ESA and Yuzhnoye, looking at future launchers and motors. Up till that point, Yuzhnoye's only involvement with ESA was through the Vega small-launcher program, where Yuzhnoye is building the RD-869 fourth stage, although as a sub-contractor for the Italian company Avio.

Employment at NPO Yuzhnoye design bureau has fallen from 10,000 to 4,500, while in the factory employment is down from 52,000 in the Soviet period to 6,000 at present. Their products go to three customers: about a third for the Ukrainian

Space Agency, a third for Russia and a third for Sea Launch, the last bringing a welcome revenue of $100m a year.

NPO PM, BUILDER OF COMSATS

The principal builder of communications satellites is NPO PM in Krasnoyarsk, located in forests on the banks of the Yenisey river. The founder was Mikhail Reshetnev as #2 branch of OKB-1 on 4th June 1959, soon renamed OKB-10. When the work of designing Russia's first communications satellites overwhelmed Korolev in the mid-1960s, he passed responsibility for the new Molniya program to him, with the center redesignated the Scientific and Production Association (NPO) Prikladnoi Mechaniki (Applied Mechanics). Little was known of NPO-PM until the 1990s, because it operated in a closed area (called Krasnoyarsk-26) and because many of its products were military communications satellites. During the Russian period, the company also began to use the civilian address of Zhelenogorsk. Having started with Molniya, NPO PM was then given the responsibility for the development of all Soviet communications satellites. The Strela series was passed there from OKB-586 in the Ukraine.

Since then NPO PM has gone on to build no fewer than 27 different space systems and over a thousand individual satellites. These included Molniya, Gorizont, Ekran and newer comsats (e.g., Ekspress), as well as the military comsats (e.g., Raduga, Strela and Gonetz). The NPO PM factory in Krasnoyarsk is a true cradle-to-grave operation, bringing each satellite through from design to fabrication, ground-testing, integration and orbital control until the mission is concluded. NPO PM had a virtual monopoly on communications satellites until the appearance of Yamal and Kupon, built by Energiya and Lavochkin, respectively, though the poor performance of both suggests the complexities of building comsats should not be underestimated.

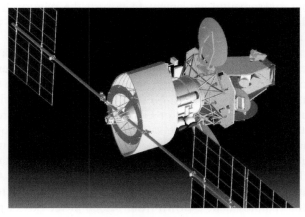

Ekspress AM-2, built by NPO PM

In its efforts to keep ahead, NPO PM brought in Western expertise from the Alcatel company. Sesat, launched by Proton in April 2000, was constructed by NPO PM in Krasnoyarsk with Alcatel's assistance for the European satellite communications organization, Eutelsat. The advanced Sesat had 18 channels and the ability to carry video, internet, mobiles, paging and software retransmission, with eight electric motors for station-keeping and a design lifetime of ten years. This became the basis of what was called the MSS767 platform of large, 2.6-tonne, 12-year, 4.2-kW, heavy, 24-hr comsats with 20 to 30 transponders and the model for the Ekspress AM series (Ekspress AM1, 2, 3).

Since then, NPO PM has developed both a larger and smaller line of comsats, the Ekspress 2000 and Ekspress 1000 lines, respectively [5]. The Ekspress 2000 is a 3.2-tonne comsat, with 25 kW of power, with up to 60 transponders and a lifetime of 15 years and suitable for launch by Proton M. Satellites launched with this platform will be called the Ekspress AT series and the Ekspress AM30 and AM40 series. The second, smaller type is the Express 1000 platform, 700 to 1,400 kg, with 10 to 12 transponders, 2 kW of power and 15-year life expectancy, with the satellites called the Ekpress AK or in its navigational version the GLONASS K. These are designed to be launched by the Soyuz Fregat or several at a time by the Proton. An even smaller platform for a 24-hr satellite, with eight transponders, is in consideration, called Gnom.

About 10,000 people now work there. Its founder, Mikhail Reshetnev, died in January 1996. Granted the size of the expanding communications market in Russia and abroad, it should be able to secure for itself a strong future.

KB ARSENAL: THE OLDEST DESIGN BUREAU

Traditionally, the main builder of military satellites was the Yuzhnoye Design Bureau in the Ukraine. Despite the importance of St. Petersburg as an industrial center in Russia, remarkably little space work is carried out there. For sheer longevity in weapons-building, the Arsenal factory is probably unequaled, for it was set up in 1711 by Tsar Peter I as a canon foundry, but became OKB-7 in 1949. From 1980, construction of the US P series of EORSATS was transferred there, and it was renamed the Arsenal Design Bureau of St. Petersburg in memory of M.V. Frunze. In the new century it took on responsibility for the replacement program for the EORSATS and the Tselina 2 electronic reconnaisance program, Liana. Arsenal can claim to be the oldest design bureau connected to space research, and it is likely to be a major supplier of military satellites in the future.

TSSKB SAMARA: CONTINUOUS PRODUCTION FROM 1957

Outside Moscow itself, the greatest concentration of rocketry in the Russian Federation may be found in Samara. Responsibility for building Korolev's R-7 rocket and later the N-1 rocket went to OKB-1's #3 branch in Kyubyshev on the river Volga in

Progress rocket factory, Samara

1959 where it took over an old aircraft factory. In 1974 the #3 branch separated from OKB-1, becoming independent as the TsSKB, or the *Tsentralnoye Spetsializorovannoye Konstruktorskoye Buro*, a sonorous title meaning Central Specialized Design Bureau. The plant actually comprises three elements: the design bureau, the huge rocket factory for the R-7 (the Progress plant or the Progress works) and an affiliate design bureau, KB Foton. To complete these naming complexities, Kyubyshev itself was since renamed Samara; and, on top of that, the original parent company, now NPO Energiya from which it had originally split, returned to Samara many years later to set a up new branch there called the Volga branch.

Originally the TsSKB was the Duks bicycle factory, set up by a German businessman Jules Muller in the Baltics in 1884, which by World War I had expanded into the production of motorcycles, cars and trolleys. Duks was nationalized in 1918 as State Aviation Plant #1, evacuated to Kyubyshev in 1941 (Goebbels prematurely trumpeted its destruction by the Luftwaffe) and during the war it turned out Ilyushin 2 dive bombers. The postwar TsSKB was the creation of Dmitri Kozlov, and it was under his guidance that the plant assumed responsibility for manufacturing the R-7, which has been in continuous production there since 1956. Korolev was obviously pleased with Kozlov's progress, for he then gave him responsibility for the Zenit satellite. Over time, TsSKB became the design center for the Yantar and subsequent spy satellites like the Orlets series. Naturally, the TsSKB also developed the associated civilian models of Zenit and Yantar, like Bion, Foton and Resurs. The Resurs DK Earth resources satellite launched in 2006 is named after its founder Dmitri Kozlov (DK), who eventually retired after leading the bureau for 45 years.

By 1999, TsSKB had turned out over 1,500 rockets and over 900 satellites in the Zenit, Yantar, Orlets, Bion, Foton and Resurs series. Up to 25,500 people work in the Progress factory. Not all of them work in the rocket or satellite sections—about 5,000 do, of whom 360 are on the R-7 production line at any time. In the spirit of post-Soviet diversification, the factory has also branched into machine tools, vodka and sweets! In 1995 it began a profitable joint venture with the French company Starsem.

Fitting shroud to Soyuz

NPO POLYOT

Moving farther away from Moscow, Omsk is the location of the Polyot Production Corporation. Although a design bureau by origin, in practice its work focuses on production of spacecraft and rockets designed elsewhere. Originally, Polyot made Tupolev bombers in Moscow, but was evacuated to Omsk, beyond the Urals, in 1941. Polyot began to manufacture rockets and spacecraft from the 1960s and became involved in the manufacture of the R-12 small Cosmos rocket and engine production for the Energiya rocket (RD-170). During the 1990s it produced the Cosmos 3M rocket and navigation satellites (Nadezhda, Parus, GLONASS and GLONASS M). Polyot's Western partner is the German company OHB. Up to 20,000 work there now.

ORGANIZATION OF THE SPACE PROGRAM

The organization of the Soviet space program required a massive managerial effort—in the case of the Moon race, not a very successful one. During the Soviet period, considerable Western intelligence effort went into identifying the organs which took decisions on Soviet space policy. Western observers made the mistake of searching for NASA-style executive agencies which would issue commands for plants to implement. That was how a command-and-control society was run; or, so they thought. In

Polyot, builder of the Cosmos 3M

reality, decision-making in the Soviet space program was diffused and, one might say, confused.

During Khrushchev's time (1957–64), decisions were taken on an *ad hoc* basis by himself, the party and government, often very informally. Not until Leonid Brezhnev came to power in 1964 was an attempt made to impose a more orderly and rational line of command and decision-making. A department of government was made responsible for the space industry, the Ministry of General Machine Building (MOM, or *Ministerstvo Obshchego Machinostroyeniye*), headed by a government minister—Sergei Afanasayev (1918–2001) during most of the Soviet period—in conjunction with the Commission on Military Industrial Issues (VPK), with key decisions being taken jointly by the party and government, taking the form of joint decrees (e.g., the key decisions on the Moon race). The Ministry of General Machine Building was a descendant of ministries responsible for the arms industries from the 1930s and the impenetrable names of the ministries made it difficult for outside agencies to identify their true purposes—the Ministry of Medium Machine Building was the cover for the nuclear industry; only the Ministry of Heavy Machine Building had a vaguely truthful title, being responsible for cranes and excavators.

The chain of command was complicated by other powerful actors—such as the Academy of Sciences—and the practice of rival design institutes appealing to ministers and officials in MOM, the VPK, the government and the central committee to change, amend and cancel decisions, and they often obliged. In the lunar and interplanetary program, the Institute for Space Research (IKI in Russian) was established by Brezhnev in 1965 to bring some order to the deep-space program—arguably it did—but it also became another power center where battles over priorities, programs and projects were fought and refought.

Not until 1985 were the space roles of MOM and the VPK brought together when Oleg Baklanov was appointed by Mikhail Gorbachev as combined Minister of General Machine Building and the head of the commission on Military Industrial Issues—a reward he reciprocated poorly by joining the *putsch* against Gorbachev in 1991. Baklanov ended up in prison. The Soviet space program did not have executive agencies and was poorly coordinated, making its achievements, despite this, all the more remarkable. Only Chief Designer (*Glavnykonstruktor*) Sergei Korolev was ever able, through his bullying and force of personality, to overcome the multi-centered organizational architecture. Chief Designer Glushko took a different approach, merging several design bureaus together under his direction and getting for himself a seat in the government.

The Academy of Sciences of the USSR was never under direct governmental control, though it chose not to directly confront or contradict Soviet government. The academy was a self-perpetuating body of learned men and women of science—about 300 full members and 300 corresponding members—that long pre-dated the revolution of 1917. It had only a limited formal role in the organization of the Soviet space program, but was influential, especially during the presidency of Mstislav Keldysh, when it was called several times to adjudge programs and projects for manned and unmanned spaceflights. After 1991 the academy was renamed RAN (the Russian Academy of Sciences).

NEW SPACE AGENCY

With the fall of the Soviet Union, the Ministry of General Machine Building disappeared. In its place Boris Yeltsin quickly created the Russian Space Agency. Now, Westerners could at last identify the NASA-type body for which they had long been searching.

The Russian Space Agency, RKA, was formed by a decree of President Yeltsin on 25 February 1992. Yuri Koptev, born in Stavropol in 1940, who had moved from designing Mars landers for Lavochkin to the now-abolished Ministry of General Machine Building, became its first director. This gray and burly administrator took over MOM's old building, but was allocated a fraction of its staff, between 200 and 300. Interestingly, a week later, on 2nd March 1992, Ukrainian President Leonid Kravchuk decreed the formation of the National Space Agency of Ukraine (NSAU), with centers in Kharkiv and Dnepropetrovsk. The Russian Space Agency, which soon developed an attractive patch and logo to rival NASA's, had nine divisions: state programs, manned projects and launch facilities, science and commercial, international, ground, external, legal, resources, and business. Later, it was renamed Roskosmos.

Nice logo or not, the RKA had none of the authority—never mind the budget—of NASA. Yuri Koptev worked hard to maintain and develop the space program.

Yuri Koptev

However, it was difficult for him to make his mark when budgets were shrinking, agreed financial allocations did not arrive and long-term planning was necessarily replaced by the month-by-month effort to survive. The powerful and far-flung design bureaus each fought their battles to protect their own spheres of operation and to survive, an art which they had refined to perfection for many years. The survival of the manned space program depended on the actions and decisions of the RKK Energiya, owner of Mir, much more than anything the Russian Space Agency might decide. The lack of RKA authority was apparent when the Buran project was canceled: it was not a decision of the agency, as one might have expected, but rather a decision of the Council of Designers. Later, the decision on whether to keep Mir in operation beyond 1999 was made by the private shareholder board of Energiya, not by Koptev, who was equivocal about the matter, nor by the government, which made its views known but was not legally able to enforce them. The decision to keep Mir flying at a time when Russia could not get its first space station modules into orbit because of lack of money enraged the Americans and they again sought out a controlling single agency, or individual, responsible, but could not find one.

Although the RKA issued plans and project lists, their relative weighting and priority was unclear and in no way matched the hard choices which the reforming Dan Goldin was imposing on his NASA colleagues at the same time. The development of the Angara rocket was, again, a function of Khrunichev's ability to attract resources rather than a conscious long-term planning decision by the RKA. This is not to be unfair to Koptev, for he was probably the right man for an impossible job at a terrible time. It was Yuri Koptev who saw the opportunity presented in the summer of 1993 to merge the Mir 2 project with the American space station and it was Koptev who steered that merger through and who kept the International Space Station on track, despite many difficulties, ever since. His peer, Dan Goldin, had much to thank him for, something he often did.

In March 2004, President Putin abruptly retired 67-year-old Yuri Koptev, his position being taken by 58-year-old General Anatoli Perminov. He had been a missile officer since 1957, commander of Plesetsk from 1991–93 and then the first commander of the Space Forces when established by President Putin in 2001.

The precise role played by the Russian political leadership in the space program has been difficult to discern. President Putin, who came into office in 2001, was always publicly supportive of the space program, emphasizing that its achievements must not be wasted. He visited space facilities early in his post, starting with a tour of Energiya, subsequently visiting Star Town, Khrunichev and the Yuzhnoye Design Bureau, having his picture taken in a white coat in the clean rooms. He appears to have met space industry directors from time to time, both individually and, for example, in April 2004, collectively. At the Zhukovsky Air Show in 2005, he made a point of visiting the Kliper space shuttle in the company of RKA director Anatoli Perminov and senior officials from the European Space Agency.

President Putin tried to achieve some form of clearer division of responsibilities and transparency between the military and civilian sides of the space program. In March 2001 a separate Space Forces command was established, ruled by a Space Forces Military Council, covering the rocket troops, military satellites (photo-reconnaissance, electronic intelligence, communications), early-warning system and military control centers. The first person in charge was General Anatoli Perminov, and when he moved to the Russian Space Agency he was replaced by Vladimir Popovkin.

RUSSIA'S SPACE BUDGET

Table 7.2 gives the figures for government investment in space programs (2006). These are the best available international figures. All present some problems and are complicated by exchange rates and quite different labor costs, especially in China's case. These figures are current running costs only and do not include historic investment in ground infrastructure, which in Russia's case is extensive. But they do include current capital investment, which in Russia may be low, but higher elsewhere,

Table 7.2. World space budgets, 2006

Country	Budget (€ billion)
United States	29.546
Europe	5.375
China	2
Russia	2
Japan	1.970
India	0.580

Source: Euroconsult, 2006

especially China. The one element which the Russian figure understates is the level of foreign payment for services, probably the highest in the world. The level of investment from Western companies through joint ventures may be considerable at this stage, even though its impact on the program is uneven, some parts benefiting significantly and others not at all. The level of foreign investment is estimated to be in the order of €600m annually. Energiya is estimated to receive half its earnings from abroad and some estimates have even been given that 74% of its earnings come from abroad.

The Russian figures do reflect the staggering fall in the level of investment in the space program in Russia, in the order of 80% over 1989–99. The Russian space program is now one of the least well funded in the world. We know that the workforce fell from 400,000 just before the end of the Soviet period to around 100,000 ten years later, where it leveled off.

Calculating Russian space budgets is a fraught exercise. During the 1990s substantial parts of the budget either arrived later or did not arrive at all. In 1995 only 77% of the agreed budget arrived, falling to 54% in 1997 and 49% in 1998: indeed, that year, the year the ruble collapsed, seems to have been the worst year for nonarrival of funding. In 1999, 63% of the budget arrived, the first improvement. From then, the money arriving gradually came back to the stated level—but even so, it was uneven. Even when payment was made, this was sometimes in the form of loans, notes and undertakings rather than "real" money. The government once paid NPO PM promissory notes on a bankrupt bank. This was not appreciated. A feature of Russian rocket launchings was the surge of launchings at year's end, followed by a relatively quiet spring, indicative of the annual budget not coming through until year's end.

Despite this difficult past history, things improved in the new century. Table 7.3 details the Russian federal space budget. There appears to have been a substantial

Table 7.3. Russian federal space budgets, 2001–06

Year	Allocation (R billion)	Notes
2001	4.8	Of which GLONASS R1.6bn
2002	9.712	GLONASS R2.7bn, ISS R6.54bn
2003		ISS R3bn
2004	12 (then €750m)	Ekspress, R3.3bn Gonetz: R150m Resurs, Monitor, Sich: 149.9m ISS: R3.88bn GLONASS: R4.95bn Foton M: R650m
2005	21.59	GLONASS: R2bn
2006	23 (about €2bn)	

real rise from 2003 to 2004, when the space budget effectively doubled, reflecting an improved flow of revenues to the state. An important feature of the increase was that funding for science resumed after a break of almost twenty years. The budget for 2007 was set at R24.4bn—but note that this is the figure for the federal, civilian and space budget and does not include the military space budget, R11bn.

Trying to calculate Russia's space budget is rather like being in Plato's cave and trying to figure out developments on the basis of shadows of the real world. From this, it is apparent that:

- Russia's space budget benefited from a real rise over time, from R4bn to R23bn, higher than the inflation rate;
- as time progressed, more of this budget was actually paid;
- there were distinctions between those portions of the space budget allocated for the military, the ISS, GLONASS and the rest;
- there was an unspecified, hard cash, Western inflow from commercial operations;
- although Russia benefited from a huge legacy of historical investment in its space program, its current operational budgets are still low. This makes its current level of activity, which outstrips all countries except the United States, all the more remarkable.

FROM COMMERCIALIZATION TO SPACE TOURISM

So, how does one explain the difficult state of Russian space finances with its ability to maintain a space program at all? The explanation lies in commercialization. The first socialist space program was forced to become, in the shortest possible time, one of the most competitive, capitalist and global in the world.

Even during the Soviet period, space planners had begun to open their space program to the West and search for commercial opportunities. Mikhail Gorbachev established a promotional agency caled Glavcosmos to sell Russian space technology to the West, especially the Proton rocket and produced glossy advertising material.

Friendly countries expecting launches from the Soviet Union now had to pay more. In 1988 the USSR had put India's first Earth resources satellite into orbit for $2m, but by the time of the second launch, just before the dissolution of the USSR, the price had risen to $14m. Even still, old habits died hard. When the financial package to put Briton Helen Sharman up to Mir collapsed in 1991, they went ahead and flew her anyway. But she got the last free, or nearly free, flight going. More or less from the time of the *putsch* onward, prices were charged out on the basis of the real cost, with a suitable profit. The cost of Soyuz seats rose to $20m, more if a significant payload package were to be carried. The European Space Agency paid double this amount for the two long-duration missions to Mir.

Privatization was an early feature of this process. The largest design bureau in the Russian space industry, Energiya, was ordered to be part-privatized by presidential decree. Of its R1bn equity (then valued at $650m at prevailing exchange rates) 49% was offered to its management, employees, citizens, institutional investors and foreign interests. In July 1998, in an effort to maintain Russia's commitments to the International Space Station and finish off building the Zvezda service module, a further 13% of Energiya was privatized, bringing in $120m, reducing the government shareholding to 25%. Overall, privatization was not the success hoped for: the companies were grossly undervalued, there was little money to invest and there was a flood of other, bigger privatizations under way at the time. Some state companies, notably Khrunichev, managed much better without privatization.

The core of the commercialization process was the establishment by the design bureaus of joint ventures and other forms of commercial arrangements with European and American companies. These joint ventures and partnerships did not necessarily meet with total approval in Russia. Some resented the way in which the rocket program's family silver has been sold off at bargain basement prices to rivals who stood to gain huge profits from their lifetime's investment. The joint ventures drew criticism that they would lead to a brain, patent and knowledge drain to the United States and that the once-great Russian rocket industry will lose its ingenuity and ability to innovate. Their colleagues counter and say that they have no choice if they are to survive. Most Russian engineers are sanguine about the prospects, one in Samara being recently overheard saying that they had coped with two revolutions (1917 and 1991) and "we're still here."

Links between Western companies and Russia were as extensive as they were complicated. A study by the Center for the Analysis of European Security compiled an inventory in 1999 of European–Russian cooperation and itemized 87 joint projects and enterprises. These ranged from rocket services (Starsem/TsSKB, Eurockot) to scientific projects (Integral) and communications satellites (Alcatel with NPO PM).

The aspect of the commercialization of the Russian space program that attracted the most public interest and attention was space tourism. Enter Dennis Tito, an

Cooperation with Europe—Foton

engineer with the Jet Propulsion Laboratory (JPL) in California who had calculated spacecraft trajectories between the planets. He had left JPL to run an investment business which had made him a multi-millionaire. He never lost his interest in space travel and when the Japanese journalist Toyohiro Akiyama flew to Mir in 1990 for a fee paid by Tokyo Broadcasting Corporation, Dennis Tito approached the Russians with his proposal for a personal, commercial mission. In the chaotic years of the mid-1990s this got nowhere, but when Russia began to look for external investment for Mir, the idea of space tourism began to appeal. Dennis Tito worked through MirCorp, the company which negotiated with Energiya for a means of prolonging the life of the Mir station and he was provisionally assigned to a flight to Mir. The going price was about $20m, a huge amount of money, but there were plenty of American and other millionaires with the money and the motive to consider the ultimate tourist destination.

Mir failed to attract the money necessary and when it was de-orbited Tito's chance seemed to have gone with it. But he would not give up and insisted on continuing his training in Star Town. So long as he was still prepared to pay for it, the Russians could hardly object, while he spent the rest of his time pressurizing the authorities to let him fly, but this time to the International Space Station.

Although there had been much discussion in American space circles about giving a greater role for private enterprise, the decision of the Russians to fly space tourists to the ISS appalled NASA when they realized, late in the day, that the Russians were in earnest. On the other side, an unusual combination of American media commentators, free enterprise advocates and space enthusiasts cheered Energiya. NASA was criticized for being small-minded about the enterprise and portraying Tito as someone who was likely to tinker with the wrong buttons and cause chaos. NASA had flown congressmen and a Saudi prince on the shuttle, so it all looked like sour grapes because someone else was doing it first.

As the flight of the first space tourist approached, NASA put pressure on Energiya to draw back from what it considered to be a playboy stunt. In a deliberately embarrassing incident, Tito was even denied access to training with his Russian colleagues in Houston, on the basis that he wasn't a "real" astronaut. His colleagues, Talgat Musabayev and Yuri Baturin, announced that they would go out on strike, but after a day's protest Tito insisted they complete their American training without him. President Putin upped the ante during his 12th April Cosmonautics Day visit to Star Town by singling out a startled Tito, greeting him and wishing him a successful flight. NASA countered by instructing its staff not to promote Tito's flight in any way or even publish his picture. Tito had to sign a waiver to pay for the damage he might cause to the orbital outpost and NASA threatened to bill Energiya the costs of any delays arising from his presence on the mission. Moreover, NASA officials insisted that he not go into their part of the station and officially he spent his time in Zvezda and Zarya. Later, they relented, but any visits to the U.S. segment had to be under escort and he could not get an export licence to use one of his cameras there.

NASA's hostility was not shared by the astronauts on board who were pleased to welcome a fellow countryman. During his visit, Tito made a point of keeping out of the way of his colleagues when they were working and helped them by preparing their

Space adventurer Greg Olsen returns

dinners. Tito did what the cosmonauts loved to do most: watch the Earth go by and film it, except that he had all day in which to do it and turned on a tape of his favorite operas while he did so. When he returned to Earth, Dennis Tito was shaky and had to be helped from his cabin, but his enthusiasm was undimmed. There could have been no better "first space tourist", for he raved about the flight afterwards and encouraged other like-minded affluent business colleagues to consider such a mission themelves. NASA's adverse comments ended up attracting attention to the mission and Tito was invited to speak to a congressional committee about the flight, where he impressed the learned members with his knowledge and observations [6].

NASA calmed down after this and subsequent tourist missions passed with little comment, with the space tourists admitted to the Houston end of the training program. Mark Shuttleworth was a South African dotcom millionaire who developed, at his own expense, an experimental package for his mission, assembling a science panel for the purpose, with the condition that the results be put into the public domain. The experiments covered such areas as embryo and stem cell development in mice and sheep, the effects of microgravity on muscles and the development of medicine to combat HIV. He carried out a program of educational work in his country's schools on his return.

The tourist missions became more formalized and came to be operated through a Virginia-based company, Space Adventures, which had built up, over the years, a program of adventure tourism in Russia, offering weightless flights on Ilyushin 76s and high-altitude flights to the edge of space on the MiG-25 aircraft. The Russians, for their part, laid down conditions for medical fitness, training, familiarization with Soyuz systems and making full payment in advance. Prospective tourists were expected to spend at least several months training with their prospective crew members, focusing on life support systems, evacuation procedures, emergencies, spacecraft engineering and how to manage housekeeping issues. Although they were expected, like the NASA astronauts flying on the Soyuz, to learn Russian, in practice they struggled with this and the cosmonauts made the effort to learn English and meet them, linguistically speaking, half way.

Throughout the following period, there were many rumors and reports of tourists and would-be tourists who entered training. Singer Lance Bass was due to fly at one stage, but he was unable to raise the final payment, so his mission was canceled. In 2006, Space Adventures offered the possibility of a spacewalk as part of the package.

Third space tourist Gregory Olsen, 60 years old at the time of his mission, was accepted for training in 2003 and assigned to a mission for the following year. Gregory Olsen, from Brooklyn, New York, was an inventor, scientist and founder of Sensors Unlimited which commercially developed infrared cameras. He sold his company for $700m in 2000, so paying the fee was the least of his problems, as he soon found out. The ever-strict Russian doctors disqualified him four months before his mission because of a query on his lung X-rays. Not one to give up, he kept going back to doctors until he got himself cleared. He threw himself back into training with a vengeance, swimming and running hard every day and taking the vestibular chair to prepare for possible space sickness. During his mission he carried out a number of experiments for the European Space Agency. According to Olsen, the term "space tourist" is not a fair one, because it belittles the amount of training required for such a mission (900 hr) and the responsible role he was given for Soyuz systems during the mission [7]. The fourth space tourist, Anousheh Ansari, marked a fresh milestone, for she was the first woman space tourist. Anousheh Ansari was a successful businesswoman from Iran who had settled in the United States. Originally, Japanese businessman Daisuke Enomoto had trained for the mission, but he was disqualified only two weeks before take-off because of a kidney problem. Anouseh Ansari had been listed as backup, but she had only six months in training and had not expected to fly until the following year. Her interest in space travel, she told the pre-flight press conference, had come from watching the science fiction series *Star Trek* and she sponsored the X-prize for the first private sub-orbital flight.

She nearly fell over when she was suddenly told by Space Adventures to get ready for a trip into orbit in only two weeks and, in line with new trends in technology, immediately opened a blog to describe the countdown and subsequent mission. Although she made no secret of her continued strong attachment to Iran, she took advice of not wearing overt political symbols or flags of the country. Once in orbit, she recalled, she couldn't stop giggling about the experience of weightlessness, but when she first looked out the window to see our blue planet spin gracefully below, the spectacle took her breath away. The journey to the space station brought her the

Anousheh Ansari preparing for mission with Michael Lopez Alegria and Mikhail Tyurin

traditional bout of space sickness—nausea, back pain and headaches—but it had cleared by the time she arrived at the ISS. Once there, she carried out four experiments for the European Space Agency, on chromosomes, bacteria, blood cells and back pain [8].

Space tourists were not put through the full extremes of survival training. Splashdown training was done in the hydrotank, rather than the Black Sea. Instead of being sent to the Arctic, they were sent to a forest not far from Moscow, where, like boy scouts or girl guides, they learned how to bivouac and make a fire. They were also taught how to use a pistol, not an academic matter, for one returning crew long ago came down in a forest and was surrounded by wolves. In case things went wrong, instructors were not far away in any case.

Space tourists to the ISS

30 Apr 2001	Soyuz TM-32	Dennis Tito
25 Apr 2002	Soyuz TM-34	Mark Shuttleworth
30 Sep 2005	Soyuz TMA-7	Greg Olsen
18 Sep 2006	Soyuz TMA-9	Anousheh Ansari

The space tourists should not be confused with agency agreements in which foreign space agencies paid for a seat on the Soyuz. Although the amount of payment was similar and although missions followed the same profile (up-and-down, week-long visit), the astronauts concerned were expected to follow a full-time scientific program. The equipment was often flown up in advance on the preceding Progress mission, adding to the cost. Such agreements dated to the 1990s, but they had run their course with the end of the Mir program. The first new such agreement was made between the Russian Space Agency (RKA) and the European Space Agency (ESA) in May 2001. At the European end, the costs were shared between ESA (as representative of all its members) and the national space agency concerned. The 2001 agreement covered six such missions. The ESA fliers were already trained as European astronauts—indeed, Pedro Duque had by now flown on the shuttle—so this was quite different from space tourism.

ESA astronauts to the ISS

Date	Spacecraft	Astronaut	Mission
21 Oct 2001	Soyuz TM-33	Claudie Hagneré (France)	Andromède
25 Apr 2002	Soyuz TM-34	Roberto Vittori (Italy)	Marco Polo
3 Nov 2002	Soyuz TMA-1	Frank de Winne (Belgium)	Odyssey
18 Oct 2003	Soyuz TMA-3	Pedro Duque (Spain)	Cervantes
19 Apr 2004	Soyuz TMA-4	André Kuipers (Netherlands)	DELTA
15 Apr 2005	Soyuz TMA-6	Roberto Vittori (Italy)	Eneid
(4 Jul 2006	Shuttle STS-121	Thomas Reiter (Germany)	Astrolab)

Brazilian astronaut to ISS

30 Mar 2006	Soyuz TMA-8	Marcos Pontes	

The ESA astronauts attended a full training program in Star Town, including survival training and a day's splashdown training in the Black Sea. The training program comprised learning Russian (many found this the hardest part), lectures on the Soyuz and ISS systems and many hours of simulator training. Their survival training involved two days and nights at two extremes: the Siberian winter in Tiksi (down to $-58°C$, logged by meteorologists as one of the coldest places on Earth)—with refresher training in forests near Moscow—and the central Asian desert in a place renowned for scorpions ($+50°C$). The Siberian training was so cold that even in an igloo it was hard to sleep. Deserts were extreme too, for even in summer, desert temperatures fell to zero at night.

Typically, the visiting European missions carried between 12 and 20 experiments. The DELTA mission, which stood for Dutch Expedition for Life, Technology and Atmospheric research) carried 21 experiments for biology, microbiology, physics, physiology and Earth observations. Roberto Vittori became the first agency astronaut to visit the space station twice. His second mission, Eneide, had individual experiments on plant germination, upper-body fatigue, the durability of electrical components for microsatellites and parts of the forthcoming European navigation system Galileo. The Eneid mission involved a total of 23 experiments in the areas of human physiology, biology and technology demonstrations.

Winter training

The Brazilian mission was one organized by the government and by the Brazilian Space Agency to commemorate the airship flights of Santos Dumont in Paris in 1906. Marcos Pontes brought up experiments with bean seeds and glow worms.

Thomas Reiter, the German astronaut, was actually part of the RKA–ESA framework agreement, with ESA paying Russia for a duration mission. Except for the period of the suspension of shuttle flights, the understanding between the United States and Russia was that the permanent crew of the shuttle would alternate between two Americans and one Russian and vice versa. With the full resumption of shuttle flights in summer 2006, Russia was due to take two places and the United States one, but Russia sold one place to the European Space Agency so that Thomas Reiter could make a long-duration mission. So, although he flew up to and down from the ISS on the shuttle, he actually filled the place of a Russian cosmonaut. Reiter had a long mission, almost six months, in the course of which he carried out an extensive program of experiments (called Astrolab) and made a 6-hr spacewalk on 3rd August 2006.

Following her mission, Claudie Hagneré was appointed Minister for Research in the French government. Her political career was less successful and she lost out in the governmental reshuffle that followed the disastrous French referendum on the European constitution.

ESA was Russia's main partner in agency-led visitor missions to the International Space Station. Comparable, but one-off, arrangements were entered into with the governments of Brazil (the mission of Marcos Pontes, 2006), Malaysia (2007) and South Korea (2008). Here, the procedure was for the national governments to run recruitment campaigns for their prospective astronauts, with a short-listed group then going to Star Town for training and the final selection of the best candidate.

Claudie Hagneré

A thousand applied for the Malaysian mission, the country's space agency eventually selecting a pilot, an Air Force dentist, a hospital doctor and, the only woman, an engineer.

Following its success in launching a number of space tourists to the station, Space Adventures decided to offer an even more staggering—and pricey—idea: lunar tourism. Space Adventures' proposal: to offer a six-day loop around the Moon in a reconstructed Zond cabin called the Soyuz K for €80m, with a first flight set for 2009. When the original plans for space tourism were put forward, they were considered a publicity-seeking stunt, but with a queue of millionaires ready to spend the money and go through the year-long training, Space Adventures had established a viable business. Maybe, forty years later than scheduled, Zond will make a round-the-Moon manned flight after all [9].

PARTICIPATION IN THE GLOBAL COMMERCIAL SPACE COMMUNITY

Russia's increasing integration into the world economy was further evident when its socialist-block economic associations were disbanded. The group for socialist cooperation in space, Intercosmos, was disbanded in 1994. Although the group for telecommunication cooperation, Intersputnik, remained in existence, at least on paper, what was more significant was that Russia joined the European telecommunications satellite organization, Eutelsat, becoming its 41st member.

One should not overstate the level of Russian isolation in space development during the Soviet period. The Soviet Union participated in the international fora on space development, such as the International Astronautical Federation (IAF) and the various United Nations organs, such as COSPAR. The Soviet Union ran cooperative programs with the United States from 1965, especially in space biology and together the two countries hosted annual conferences on lunar and planetary exploration. The results of Soviet missions were published in the standard international journals. The Soviet Union ran bilateral cooperation programs with a number of individual countries, notably France. At the end of the Soviet period, projects like VEGA involved a broad range of international partners.

What was different in the Russian period was the participation of Russia in the international, global, capitalist rocket and satellite business. This was a difficult process, for the United States presented a number of hurdles to such participation. First, there were security concerns about Russia benefiting from Western technology transfer that would be used in its military programs. Second, the United States feared that cheaper Russian rockets would undermine its own rocket manufacturers.

As an expression of these concerns, satellite manufacturers flying on Russian rockets were obliged to obtain an export licence: for example, for INMARSAT to fly a satellite on a Proton in 1992 presidential approval was required. Even once given, Western satellites en route to Proton and other launchers were kept under American supervision. As regards cost, the Americans imposed quotas on the number of satellites that the Russian Federation was permitted to launch, so as—like the terminology said—"to promote disciplined participation in the market." In September 1993 this was set at nine launches between then and 2001, with not more than two in any one year, with Russian prices being not less than 7% below the world average. In 1996 this was raised to 20 launches and not less than 15% below world average. This was subsequently raised to 23 and the quotas eventually expired.

Proton's first commercial Western launch took place in April 1996 when it lofted a Luxembourg communications satellite called Astra 1F into orbit. Over the next few years, Proton was to win a significant number of commercial contracts, keeping its production line open, attracting in new investment and earning profits for Lockheed Khrunichev. By summer 2000 the Proton had made 17 commercial Western launches. By 2006 Russia claimed to have the largest share of the international launcher market, 45%.

These were large payloads and big earners. At the other extreme were small satellites. A typical foreign microsatellite customer was Rubin, the German word for "ruby", developed by the company OHB in Bremen. OHB developed a standard microsatellite bus—5 kg in weight, $600 \, mm^3$ in volume—which could orbit either as a free-flier or remain attached to the upper stage, normally a Cosmos 3M. The standard bus provided attitude control, batteries and a communications system of e-mailing messages back to Earth via the Orbcomm satellite network. Different experiments were flown on each Rubin: the fifth Rubin, for example, carried a detector to measure meteoroids and space debris [10].

Genesis, launched for Bigelow Aerospace

Estimated prices of flights on Russian rockets (€ million)

Angara	25–100*
Baiterek (Angara 3)	50
Cosmos 3M	12
Dnepr	10
Proton	50
Proton M	75
Rockot	15
Shtil	0.5
Soyuz	25–30*
START	8
Strela	5
Tsyklon	20
Zenit 3 Land Launch	35
Zenit 3SL Sea Launch	60

Depends on version. Source: ESD [11].

Commercialization was a two-way channel and could also work in reverse. When the independent Russian media company MediaMost wanted to launch its communication satellite Bonum in 1998, it did not go domestic for either its satellite or the launcher, opting for Hughes to build the satellite and a Delta 2 for launcher. MediaMost came under sharp criticism from the government for its lack of patriotism for not choosing either Russian satellite builders or launchers. MediaMost, which has been considered critical of the government, retorted by pointing out that American satellites were better and that to launch them on a Russian rocket would have involved long bureaucratic delays and heavy import taxes.

As may be seen, most of the commercial launches flew on the Proton, followed by the Zenit 3SL, but a number also flew on the other members of the launch fleet.

Commercial, semi-commercial and agency launches, 2000–6, Proton K with DM upper stage

1 Jul 2000	Sirius 1
5 Sep 2000	Sirius 2
2 Oct 2000	GE-1A
22 Oct 2000	GE-6
30 Nov 2000	Sirius 3
15 May 2001	Panamsat 10
16 Jun 2001	Astra 2C
30 Mar 2002	Intelsat 9
8 May 2002	Direct TV 5
22 Aug 2002	Echostar 8
17 Oct 2002	Integral (Europe)
17 Jun 2006	Kazsat

All from Baikonour

Commercial launches 2000–6, Proton M with Briz M

30 Dec 2002	Nimiq 2
7 Jun 2003	AMS-9
16 Mar 2004	Eutelsat W3A
18 Jun 2004	Intelsat 10
4 Aug 2004	Amazonas
14 Oct 2004	AMC-15
3 Feb 2005	AMC-12
22 May 2005	Direct TV 8
9 Sep 2005	Anik F1R
29 Dec 2005	AMC-23
1 Mar 2006	Arabsat 4A (fail)
4 Aug 2006	Hotbird 8
8 Nov 2006	Arabsat 4B
12 Dec 2006	Measat 3

All from Baikonour

Commercial launches, 2000–6, Zenit 3SL (*Odyssey* platform)

28 Jul 2000	Panamsat 9
21 Oct 2000	Thuraya
19 Mar 2001	XM-2 Roll
15 Jun 2002	Galaxy 3C
10 Jun 2003	Thuraya 2
8 Aug 2003	Echostar 9
30 Sep 2003	Galaxy 13
11 Jan 2004	Telstar 14/Estrela do Sul
4 May 2004	DirecTV 7S
29 Jun 2004	Apstar 5/Telstar 18
28 Feb 2005	XM-3 Rhythm
26 Apr 2005	Spaceway 1
23 Jun 2005	Telstar 8
8 Nov 2005	Inmarsat 4
15 Feb 2006	Echostar 10
12 Apr 2006	JCSAT 9
18 Jun 2006	Galaxy 16
22 Aug 2006	Koreasat 5
26 Oct 2006	XM-4 Blues

Commercial, semi-commercial and agency launches, 2000–6, Soyuz Fregat

16 Jul 2000	Cluster 1 (ESA)
9 Aug 2000	Cluster 2 (ESA)
28 Dec 2003	Amos 2 (Israel)

Both from Baikonour

Commercial, semi-commercial and agency launches, 2000–6, Soyuz FG Fregat

2 Jun 2003   Mars Express (Europe)
9 Nov 2005   Venus Express (Europe)
13 Aug 2005   Galaxy 14
28 Dec 2005   Giove (Europe)
All from Baikonour

Commercial, semi-commercial and agency launches, 2000–6, Soyuz 2

19 Oct 2006   Metop A (Europe)
27 Dec 2006   COROT (Europe)
Both from Baikonour

Commercial, semi-commercial and agency launches, 2000–6, Dnepr

27 Sep 2000   Megsat (Italy)
                 Unisat (Italy)
                 Saudisat 1A (Saudi Arabia)
                 Saudisat 1B (Saudi Arabia)
                 Tiungsat (Malaysia)
20 Dec 2002   Unisat 2 (Italy)
                 Latinsat A (Argentina)
                 Latinsat B (Argentina)
                 Saudisat 1C (Saudi Arabia)
                 Rubin 2 (Germany)
                 Trailblazer (U.S.)
29 Jun 2004   Demeter (France)
                 Saudicomsat (Saudi Arabia)
                 Saudicomsat 2 (Saudi Arabia)
                 Saudisat 2 (Saudi Arabia)
                 Latinsat C (SpaceQuest U.S.)
                 Latinsat D (SpaceQuest U.S.)
                 Amsat Echo (amateur radio)
                 Unisat 3 (Italy)
24 Aug 2005   OICETS (Japan)
                 INDEX
12 Jul 2006   Genesis 1
26 Jul 2006   Belka and 16 small satellites (fail)
All from Baikonour, except Genesis from Dombarovska

Commercial, semi-commercial and agency launches, 2000–6, Cosmos 3M

28 Jun 2000   Tsinghua (China)
                 SNAP 1 (Britain)
15 Jul 2000   Champ (Germany)
                 Rubin 1 (Germany)
                 Mita (Italy)
28 Nov 2002   Mozhayets 3 (Russia)
                 Alsat 1 (Algeria)
                 Rubin 3 (Germany)

27 Sep 2003	Mozhayets 4 (Russia)
	Nigeriasat 1 (Nigeria)
	KaistSat (South Korea)
	UK-DMC (Britain)
	Bilsat (Turkey)
	Laretz (Russia)
27 Oct 2005	Mozhayets 5 (Russia)
	Sinah 1 (Iran)
	China DMC (China)
	NCube (Norway)
	UWE (Germany)
	XI-V (Japan)
	SSET Express (Europe)
	Topsat (Britain)
19 Dec 2006	SAR Lupe (Germany)

All from Plesetsk

Commercial, semi-commercial and agency launches, 2000–6, Rockot

17 Mar 2002	GRACE (U.S./Germany)
20 Jun 2002	Iridium (U.S.)
30 Jun 2003	Mimosa (Cz)
	DYU (Denmark)
	AAU (Denmark)
	CUTE (Japan)
	Quakesat U.S.
	MOST (Canada)
	CanX (Canada)
	CubeSat (Japan)
30 Oct 2003	SERVIS (Japan)
8 Oct 2005	Cryosat (ESA) (fail)
28 Jul 2006	Kompsat (Korea)

All from Plesetsk

Commercial, semi-commercial and agency launches, 2000–6, START-1

5 Dec 2000	Eros A (Israel)
20 Feb 2001	Odin (Sweden)
28 Apr 2006	Eros B (Israel)

All from Svobodny

What kinds of satellites were they? Most of the communications satellites launched were large, 15 year lifetime satellites carrying television, telephone and direct broadcast services for lease by media and telecommuncations companies. One of the most unusual was the XM series, XM-1, 2, 3 and 4, called *Rock*, *Roll*, *Rhythm* and *Blues*. These transmitted radio, rather that television, relaying 130 channels to car-based subscribers paying an annual fee. In effect, they re-broadcast local channels nationwide, covering a range from country music to opera, comedy to talk shows, children's programs to weather stations.

As for the other payloads, the 27th October 2005 launch typified the varied range of small payloads carried. SSET was a 62-kg technology demonstrator built by students in 23 universities, while the Chinese satellite was one of a worldwide series of small satellites built to permit communications in disaster areas. XI-V and UWE were 1-kg picosatellites, the latter built by the University of Wurzburg. Topsat was a British defense imaging satellite designed to test the sending of pictures to small terminals.

The 30th June 2003 launch on Rockot carried piggybacks with the Monitor E model. Mimosa was a 51-kg atmospheric density satellite, equipped with micro-accelerators, for the Czech Academy of Sciences, while 66-kg MOST was a Canadian telescope designed to find planets in orbit around others stars. The others were 1-kg picosatellites: DTU (a Danish tether test); CUTE and CubeSat (demonstrators for the Tokyo Institute of Technology and University of Tokyo, respectively); Quakesat, an American earthquake detector; AAU, a Danish imaging satellite; CanX a Toronto, Canada, imaging satellite.

The Nadezhda launch of 28th June 2000 carried two more satellites built by Surrey Satellite Technology Ltd. (SSTL), the British company which specialized in the construction of small Earth satellites. The first, Tsinghua, was a microsatellite built for the Tsinghua University in Beijing, China, with a 39 m resolution multi-spectral imaging camera designed to help in disaster monitoring and with a store-and-forward communications system. SNAP 1, a 1.5-m project, which was even smaller and weighing only 6.5 kg, was designed to test the communications and rendezvous

Microsatellites prepared for launch, Plesetsk

Making satellites—the shop floor

capacities of very small satellites. Its tiny size packed a GPS navigation system, camera, computer, propulsion and attitude control system. The camera was designed to look through clouds and take clear pictures of the Earth from 500 km. The computer and propulsion system was intended to enable SNAP to rendezvous and fly in formation with its companion, Tsinghua, and test out the possibility of building nanosatellite constellations.

The September 2000 Dnepr showed the type of mixture of small satellites that could be carried. Megsat was a 56-kg Italian environmental satellite, accompanied by Unisat, a 10-kg student satellite. The two 10-kg Saudisats were built by radio amateurs in Riyadh. Tiungsat, named after a famous bird, was a 52-kg remote-sensing satellite for the Ministry for Science and Technology in Malaysia.

The assembly of a global disaster warning system was a feature of a number of launches. SSTL built a series of small satellites designed to enable governmental and non-governmental organizations to better respond to natural disasters, called the DMC (Disaster Monitoring Constellation). The satellites had the ability to scan 600 km^2 with a resolution of 32 m every 24 hr, designed to watch the state of floods and earthquake damage. The satellites in the constellation were Alsat (the first for Algeria), Nigeriasat (Nigeria), Bilsat (Turkey), UK DMC (Britain) and China DMC. The September 2003 Cosmos 3M carried the largest single group of the DMC constellation, with the British, Nigerian and Turkish satellites. Also on the launch were a South Korean ultraviolet telescope and a Russian radar calibration satellite called Laretz as well as the main payload, the Mozhayets 4.

Champ was a German satellite built in Potsdam to make precise gravity field and magnetic measurements over five years using a magnetometer, accelerometer, laser reflector and global positioning receiver. Along for the ride were an Italian mini-satellite Mita and a German mini-satellite Rubin for internet users. GRACE (Gravity Recovery And Climate Experiment) was an experiment developed by the American and German space agencies, NASA and DLR, respectively, involving two identical satellites orbiting precisely 220 km apart to compile the most detailed ever space-based gravity map. NASA paid Eurockot $10m for the launch.

Integral was an important scientific mission, for it was Europe's most ambitious gamma ray astronomical mission for some time and had been many years in preparation, costing €600m. Integral required a powerful rocket to place it in its unusual eccentric orbit of 685 × 153,000 km, period 72 hr. Many ESA personnel gathered in Baikonour to watch the 294th Proton launch. It was a clear day with blue skies and they were able to watch a perfect launch, Proton making three onion-ring-shaped shock waves as it headed into the upper atmosphere.

OICETS (Optical Inter Orbit Communications Engineering Test Satellite) and INDEX (INnovative technology Demonstration EXperimental Satellite) were Japanese technology demonstrators, OICETS designed to test out the possibilites of one satellite communicating large volumes of information with another by means of laser beams. In the Japanese tradition they were renamed once they reached orbit as Kirari and Rimei, respectively.

Some of the satellites were even military ones for the West—a development unimaginable twenty years ago. As far back as 1995 the Russians had carried a small American military satellite, called Skipper, and, as already noted, put into orbit military Israeli satellites. In December 2006 a Cosmos 3M orbited a German military surveillance satellite, SAR Lupe. Circling the Earth at 500 km at 80°, its objective was to transmit 30 daily 1 m resolution radar images to the German military control center in Gelsdorf. SAR Lupe's 3 m long radar beam required a modification to the Cosmos 3M rocket and the winter launching went perfectly.

COOPERATION: ROGUE STATES

There were two sides to the technology transfer issue noted above. First, there was, as already noted, an American concern that Russia would lift Western technologies by copying the features of satellites passing through en route to a Baikonour or other launch pad. Inspecting other countries' satellites was not that difficult, as the Americans themselves knew, for in 1960 they had kidnapped, overnight, a Soviet lunar spacecraft en route to an exhibition in Mexico, disassembled it and put it back together again without anyone knowing a thing! Granted that the Russians had their own, capable communication satellites, it is not certain whether the incentives for the Russians to spy on satellites on the way to the pad were that strong. What about Russian transfer outward?

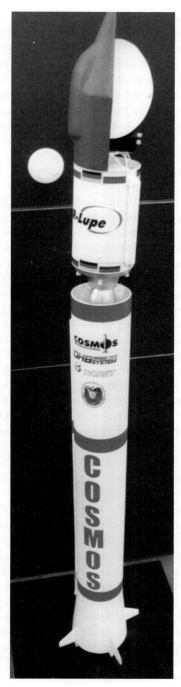

SAR Lupe

The relationship between the Russian space program and what were called in the West "rogue states" was the source of tension between Russia and the United States throughout the post-Soviet period. Essentially, this revolved around a fear that the Russian Federation, or its companies, would provide space-related assistance to the enemies of the United States, principally Iran but potentially North Korea or other countries named by the Bush presidency as members of the "axis of evil". Iran was the main focus of tension. In one incident, jumpy technology transfer officers objected to Russia supplying Iran with a fire-fighting pump on the basis that the Iranians could reengineer the technology into a rocket engine. Energomash, in particular, complained that it had never transferred as much as a bolt or a rivet to a rogue state, yet a licensing agreement with the Americans was held up for 400 days while bureaucrats picked the company apart.

The principal point of contact between Russia and Iran was the Zohreh project, the idea of an Iranian geosynchronous communications satellite. The concept dated to the time of the Shah's "white revolution" to modernize Iran in the 1970s, when it was first proposed. The satellite was named Zohreh (Farsi for Venus), the name still attached to the project despite all the political upheavals that followed. Zohreh was suspended following the Islamic revolution, but the project was revived in the 1990s. Following the failure of negotiations with France, Iran turned to Russia in 1998, but talks made only peristaltic progress. On the initiative of the Russian export board, a delegation of telecommunications experts traveled from Iran to Krasnoyarsk in May 2001 to discuss with NPO PM the possibility of the company building Zohreh. Agreement was eventually reached in 2005 for NPO PM to build two satellites in a deal worth about €100m, with some components supplied by French and German companies (Alcatel in the French case). Zohreh would be modeled on the Ekspress satellite and have up to 16 transponders for television, radio and data transmission. The agreement did not end the uncertainty over the project, which became caught up in the stand-off between Iran and the United Nations over Iranian nuclear proliferation, leading to reports that the project was delayed.

Amid all the rumors, the only concrete outcome of the Iranian–Russian connection was that in October 2005 a Cosmos 3M put up the 170-kg Sinah 1 from Plesetsk. Reports on the mission were murky, most suggesting that it was a store-and-forward communications satellite, others that it also had 50 m and 250 m resolution cameras for Earth observations (some military in nature), yet others that it was a "training satellite" for ground control. The Russian side appears to have been the Polyot company in Omsk and the Optek company, which may have furnished the optics. On the Iranian side, the organizations involved were the Electronic Industries Company, the Ministry of Science and Technology, and the Institute of Topographic Engineering. When asked to identify the purposes of the satellite, Iranian officials suggested "natural disasters, agriculture, earthquakes and natural resources." Although a photograph of Sinah was taken in Polyot, the detail was poor and it was uninformative. The Iranians themselves did not publicize any outcomes from the mission.

Another Iranian satellite, called Mesbah (Farsi for lantern), was understood to have been built at the same time: it was a smaller, 75-kg, store-and-forward communications satellite. Mesbah's builder was alternately reported to be the Carlo Gavazzi company in Italy and the Polyot bureau in Omsk [12]. This was not as contradictory as it sounded, for Carlo Gavazzi had a cooperation program with Polyot. Mesbah reportedly arrived in Plesetsk alongside Sinah for the October 2005 launching, but it was damaged in pre-launch handling and its power system could not be repaired. The launching of Sinah and other payloads went ahead without Mesbah, which was sent back to the shop. In October 2006 Iranian TV ran a program on Mesbah, wanting to know what had happened to the missing satellite, but the Iranian State Space Organization stayed silent and so did the Russians.

During the mid-1990s, there were sporadic reports that Russia was selling rocket parts, even entire missiles, to Iran, Iraq and North Korea. These appeared to be confirmed when, in August 1998, North Korea astonished the world by announcing that it had placed its first Earth satellite in orbit, though the claim was challenged in the West and few except the North Koreans ever saw the satellite or picked up its signals. At around the same time, Iraq was reported to be developing a long-range missile, one also capable to putting a small satellite into orbit, called the Al Abbas. During heated congressional hearings on Mir in autumn 1997, Congressman Weldron theatrically produced a Russian accelerometer and gyroscope retrieved from a river in Iraq where they had been dumped by smugglers: "Why are we using American dollars to fund programs that are leaking technology to America's enemies?" he asked. In June 2000 the American National Security Agency again accused Russia of selling parts and providing technical advice to rogue states. A late 1999 Central Intelligence Agency report leaked in autumn 2000 alleged that Russian companies were continuing to supply ballistic missile technology to Iran.

North Korean leader Kim Jong Il visited Russian space facilities in August 2001 on his first visit to the country. Eschewing faster methods of communication, he traveled on the trans-Siberian railway, taking a full week to reach Moscow; but once there he toured the Khrunichev factory to see Protons and Rockots and went on to TsUP mission control. Fearing demonstrators, guards were posted every 50 m at the Khrunichev plant and there was high security everywhere. He had plenty of time to reflect on what he saw, for he traveled on to St. Petersburg, whence he took the train on a ten-day route back all the way to Pyongyang via Novosibirsk. Russia offered to launch North Korean satellites—for a fee and if North Korea stopped its missile program—but North Korea seemed unwilling to pay; so little seems to have come from the visit. Russia had more commercial success and attracted less international attention for its dealings with South Korea, launching its satellites on the Dnepr and Rockot (Kompsat) and contributing to the planned national launcher, the KSLV-1. Here, the Khrunichev company made an agreement with South Korea in 2004 to assist in the development of a light launch vehicle based on the Angara 1.1 and able to put 100-kg cargoes into orbit.

It is very difficult to establish whether Russia has actively aided developing countries in their attempts to build rockets which can launch intercontinental ballistic missiles or put satellites into orbit. Four countries—North Korea, Iraq, Pakistan and Syria—developed long-range versions of the Scud missile, which is the old R-11

South Korea's Kompsat

rocket developed by the Korolev and Makeev bureau from 1953 to 1959 and subsequently exported long before anti-proliferation restrictions were agreed. It is just possible, using upper stages, to modify the Scud in such a way that it can put a very small satellite into orbit, as the North Koreans undoubtedly tried to do in 1998 and again in 2006. However, there is a considerable difference between exporting missiles in the 1960s and the government actively aiding these countries now to build up an intercontinental ballistic missile capacity. A nuclear warhead would, by definition, be a heavy object and far beyond the lifting capacities of a Scud's intercontinental range.

One country where technological cooperation is proven and well documented is China. This has, without question, earned additional resources for the Russians.

COOPERATION: CHINA

Although China was not formally a member of the axis of evil, American–Chinese relationships were poor in the early years of the new century, with the United States very wary of Chinese military ambitions and intentions. The same licensing regime as applied to Russia was also applied to China—except that licences to China were normally refused, in effect denying China the possibility of flying satellites on a commercial basis. The United States kept a close watch on Russian–Chinese technical cooperation.

There was close cooperation between the Soviet Union and China during the 1950s. The USSR gave the Chinese two German A-4s and invited Chinese students in groups of 50 to study Soviet rockets. They were sent packing when Khrushchev and Mao Zedong split in 1960. Not until early 1992 were working relations restored and the Chinese returned to Moscow. Chinese engineers and scientists came to Moscow to study the Soyuz spacecraft, Russian ground and tracking facilities and the environmental control systems for manned spacecraft. In spring 1995 Russia and China reached an intergovernmental agreement on space cooperation, specifying Russian assistance to China in the area of manned spaceflight. The Chinese bought a number of hardware items, principally a docking system and environmental control systems, both for hard currency. Two cosmonaut instructors, Wu Tse and Li Tsinlung, underwent a two year long training course in the Yuri Gagarin Cosmonaut Training Center. At the conclusion of their training, they left Star Town on 19th November 1998 to join China's recently-formed squad of astronauts, or *yuhangyuan* in Chinese.

A round of discussions took place between Russia and China in early 2000 on the future of their cooperation in manned space projects, with Russian Deputy Premier Illya Klebanov visiting China for this purpose. Russia sounded out China's interest in the Mir space station and offered China the use of Mir for €588m for a three to four-year period, venturing that this was a bargain compared with the cost of developing their own station, which would cost them over €1.1bn. Press speculation notwithstanding, Russia never considered selling Mir outright—the term was "joint utilization". In the event, the Chinese preferred to build their own station later—but the Russians were not sent away empty-handed. The two sides reached agreement for the following:

- Russia to provide technical assistance in the design of the Chinese space station;
- Russia would build a limited number of components for the station;
- training would be provided for yuhangyuan and ground controllers;
- Russia would transfer 36 specific areas of space station technology.

Two Russian cosmonauts duly went to Beijing in 2000 to provide technical consultancy for the Chinese: Anatoli Berezovoi and Anatoli Filipchenko.

When President Vladimir Putin visited President Jiang Zemin in 2002, he brought with him Russian Space Agency director Yuri Koptev. Their agreements were formalized at the Chinese–Russian Council on Space Exploration which met on 22nd August 2002 and confirmed 21 areas of cooperation. One concrete item was that the Russians agreed to fly Chinese experiments to the Russian modules on the International Space Station—a kilo of rice seeds from Harbin Institute of Technology on a six month long astroculture experiment. Soon after, it was learned that deep-space missions to the Moon and Mars had been added to the subject areas of bilateral cooperation. Soon after, the Chinese announced their plans for a three-part program of lunar exploration: an orbiter (2007), rover (2011) and sample return mission (2017). With sample return missions, Russia had a unique technology, which had to be of especial interest to them.

The Shenzhou

China made its first flight of a manned spacecraft prototype in 1999, the Shenzhou. The first manned flight took place in 2003 (Shenzhou 5, manned by Yang Liwei), followed by a week-long mission two years later (Shenzhou 6, Fei Junlong and Nie Haisheng). Shenzhou resembled Soyuz, as did the upper part of its Long March 2F launcher which had a shroud and escape system quite similar to Soyuz. In

The Soyuz spacecraft

the West, it was alleged that China had simply copied the Soyuz. Such comment ignored the manner in which the Chinese had always worked hard to develop their own indigenous technology. Yuri Koptev described the 1995 agreement as one in which Russia helped China "fill in some of the gaps" and this may be a useful analysis. Closer examination reveals significant differences between Soyuz and Shenzhou, the Chinese spacecraft being larger, with a different system of solar panels and able to detach the orbital module for an independent research program.

Discussions began in Beijing in November 2005 on a new cooperation agreement between Russia and China. The deputy director of the Russian Space Agency (RKA) led the delegation which discussed an accord to run from 2007 to 2016. Specifically, it covered human and Mars exploration (China was already planning a small Mars probe, called the Huowei). The discussions were then reported to a meeting of the Russian and Chinese prime ministers, Mikhail Fradkov and Wen Jinbao, the following week. They then signed a ten-year cooperation agreement to cover the 2007–2016 period in ten project areas. Details were sketchy, but among the cooperation areas envisaged were:

- Chinese collaboration on Mars missions;
- spacewalk training for Chinese *yuhangyuan* in Star Town;
- Russian advice to China on lunar landing probes.

One of the most interesting projects began to harden up in 2006, when it was learned that on arrival in Mars orbit the Russian Phobos Grunt probe would deploy a small Chinese sub-satellite, the Yinghuo 1, into an $800 \times 80{,}000$-km equatorial orbit. Yinghuo, weighing 120 kg, would be the first sub-satellite of Mars and would study Mars' atmosphere and ionosphere with Phobos Grunt. This was a clever concept, getting China to Mars much sooner than would have otherwise been the case, opening up the scope for simultaneous observations by two satellites and furnishing some funding for Russia.

Cooperation was extended at the 10th bilateral conference in Shanghai in November 2006. China agreed to contribute some of the soil-sampling gear for the Phobos Grunt lander, while Russia would contribute to China's planned lunar soil recovery mission planned for 2011. China also indicated an interest in taking part in the planned Mars sample return mission, Mars Grunt. By this time the two countries were cooperating on no fewer than 38 joint projects. Russia was always adamant that cooperation did not include the Kliper or related space shuttle technologies—presumably to signal that the Europeans were their favored partners in this area of endeavor.

COOPERATION: INDIA

The other country of concern to the United States was India. The Indian case was somewhat different, for India did not raise the same military apprehensions as China and played a much less aggressive role in global capitalism than did the Chinese. Despite this, Russian assistance to India was a cause of tension with the United States.

This arose originally in the 1980s, when India moved toward building a large rocket able to put both national and commercial satellites into 24-hr orbit from its near-equatorial launching base in Sriharikota, near Chennai (Madras). This launch vehicle, later called the GSLV (GeoSynchronous Launch Vehicle), required a high-powered upper stage, preferably using hydrogen fuels, something far beyond India's capacity at that time. India especially wanted to orbit Gramsat (later called Gsat), a multi-purpose telecommunications satellite to bring educational television to the villages of rural India. Knowing it would take at least 10 to 15 years to develop the technology itself, India sought help from abroad, making approaches to Japan, the United States, Europe and, when these came to nothing, Russia.

Here, the Isayev Design Bureau offered its KVD-1 engine. No one had heard of the engine, for the West was certain that the USSR had never been able to develop cryogenic engine technology. The KVD-1 engine has a thrust of 7,100 kg, a specific impulse of 462 sec, a burn time of 800 sec and a combustion chamber pressure of 54.6 atmospheres. It was 282 kg in weight, 2.1 m tall and had a diameter of 1.6 m. The KVD-1 used a turbopump-operated engine with a single fixed-thrust chamber; its two gimbaled thrust engines could operate for up to 7.5 hours and be restarted five times. The KVD stage weighed 3.4 tonnes empty and 19 tonnes fueled. The KVD-1 was first test-fired in June 1967 and was tested for 24,000 sec in six starts; its thrust and capabilities were unmatched for years. But the KVD-1 never flew and when the manned lunar program was finally canceled in 1976 it went into storage.

The Indians made an agreement for the GSLV to fly the KVD-1 as the upper stage, but this was denounced by President George Bush Sr. as a violation of the Missile Technology Control Regime and he applied commercial sanctions on India. His successor, President Clinton, offered to withdraw them if the Russians transferred

KVD engine

individual engines, but not the production technology that would enable India to design its own cryogenic engines. The Indian negotiations soon became caught up in the negotiations for the joint Russian–American plan to build the International Space Station taking place at that very moment. Russia insisted that if it pulled out of the technology transfer deal with India there must be compensation, which the Russians valued at $400m. This became the exact price which the Americans paid the Russians for the seven American flights to Mir.

The Indians of course were furious at Russia for pulling out of the agreement. Russia and India negotiated a fresh agreement in 1994, whereby Russia agreed to transfer three, later renegotiated by the Indians to seven, KVD-1s intact, without the associated technology, all for a price of $9m. Only two KVD-1s would fly, after which the Indians would develop their own upper stage. The negotiations were later described as very tough, but the Indians managed to negotiate a lower price and extra engines in exchange for the loss of technology transfer. India was required to satisfy the United States that it would only use the KVD-1 purely for peaceful purposes, not to re-export it, nor to modernize it without Russia's consent. The Indian version was called the KVD-1M.

The first of six KVD-1Ms duly arrived on 23rd September 1998, and one powered India's maiden GSLV launch on 18th April 2001. Disappointingly, the KVD-1 underperformed, burning for 706 sec instead of the 710 sec planned, putting Gsat 1 into a geosynchronous transfer orbit, but one lower than planned (32,000 km instead of 36,000 km). Despite this discouraging start, two subsequent launches were successful. Although there had been an understanding that the KVD-1M would be used only on the first two launches, it was used on the next two as well. This may have been because the indigenous Indian stage was not yet ready, but the Indians may have taken the view that, having paid for the engines, they were going to use them.

Relationships with the Americans improved in the intervening period, President Bush Jr. making a successful visit in 2006; the Americans may have decided not to pursue the matter further.

There is good reason to believe that the Indians managed to get the blueprints in any case in the course of four shipments from Moscow to Delhi on covert flights by Ural Airlines. The Indians argued that the KVD-1 had nothing to do with missile technology control at all, since the KVD-1 would be of little value as a weapon against their most likely enemy, their neighbour Pakistan, which it could already bombard with short-range missiles. According to the Indians, the episode actually had more to do with keeping India out of the competitive global 24 hr satellite launching business.

GSLV launches with KVD-1M upper stage

18 Apr 2001	Gsat-1
8 May 2003	Gsat
20 Sep 2004	Edusat
8 Jul 2006 (fail)	Insat 4C

All from Sriharikota

ORGANIZATION: CONCLUSIONS

For all this endeavor, what was the outcome? Here we look at the launch rate of the Russian space program (Table 7.4). The contraction of Russian space activity was very evident if we consider that the Soviet-period peak of 102 launches in 1982 fell to 23 in 1996. One must be cautious, nonetheless, for launch rates had already begun to fall, even in the Soviet period. This was due to satellites operating for longer, so there was less need to replace them so speedily. The Zenit first generation of photo-reconnaissance satellites, which orbited for only a few weeks, gave way to more capable Yantar digital satellites, able to operate for up to a year. Communications satellites, formerly guaranteed for only 2–3 years of operation, were now built to

Table 7.4. World launch rates, 2000–06

	2000	2001	2002	2003	2004	2005	2006
Russia	36	25	25	24	26	27	29
U.S.A.	28	23	18	23	18	13	18
Europe	12	8	11	4	3	5	5
China	5	1	4	6	8	5	6
Japan		1	3	2		2	6
India		2	1	2	1	1	
Israel			1				
Total	*81*	*60*	*63*	*61*	*56*	*53*	*63*

operate up to twelve years. In other words, the high launch rates of the 1980s would have fallen in any case.

Overall, the program survived the transition of the Russian period. The design bureaus managed to retain their position. There was a rapid rate of commercialization, joint enterprises, space tourism and cooperation with other countries which brought in fresh resources. The federal space budget, although small compared with that of other countries, provided a rising base on which future activities could be developed (Chapter 8). In 2000 Russia's position in the world launching league began to improve and Russia moved back into top slot, which it has held every since, maintaining an annual launch rate in the mid-twenties ever since.

REFERENCES

[1] For a history of the evolution of the design bureaus, see William P. Barry: *The missile design bureaux and Soviet manned space policy, 1953–1970*. Doctoral thesis, Merton College, University of Oxford, 1996.

[2] Oberg, Jim: Pod people. *Air & Space*, October/November 2003.

[3] Prisniakov, V.G.; Kavelin, S.S. and Platonov, V.P.: *Sources of Ukrainian space potential on the 85th anniversary of V.M. Kovtunenko*. Paper presented to the International Astronautical Federation, October 2006.

[4] Lardier, Christian: Youjnoe mise sur l'Europe. *Air & Cosmos*, #1987, 10 juin 2005.

[5] Lardier, Christian: Les futurs satcoms du russe NPO PM. *Air & Cosmos*, #1882, 21 mars 2003.

[6] French, Francis: Citizen explorer—the Byzantine odyssey of Dennis Tito. *Spaceflight*, vol. 44, no. 4, April 2002; Vis, Bert: Space station tourists—an astronaut's perspective on Tito's controversial first. *Spaceflight*, vol. 46, no. 11, November 2004.

[7] Da Costa, Neil: A private trip into space: Gregory Olsen—the third "space flight participant". *Spaceflight*, vol. 48, no. 2, February 2006.

[8] Quine, Tony: This beautiful planet—last-minute trip to space station for Ansari. *Spaceflight*, vol. 48, November 2006.

[9] Jha, Alok: Fly me to the moon—and let me pay among the stars. *The Guardian*, 12th August 2005.

[10] Ziegler, Bent; Kalnins, Indulis; Bruhn, Feredrik and Stenmark, Lars: *Rubin—a frequent flier testbed for micro and nano-technologies*. Paper presented to International Astronautical Federation Conference, Valencia, Spain, October 2006.

[11] ESD: *European Space Directory*, 2006. 21st edition. Paris, ESD partners, 2006.

[12] Pirard, Théo: L'Iran bientôt acteur dans l'espace. *Air & Cosmos*, #1996, 9 septembre 2005.

8

Resurgent—the new projects

As we saw in Chapter 1, the Soviet space program in the 1980s had been the admiration of the world. *The USSR in outer space: the year 2005* promised a huge orbital station serviced by a large space shuttle. Applications satellites circled the Earth while deep-space probes set out for distant destinations. Rovers roamed the plains of Mars to bring samples to rockets that fired their cargoes back to Earth. Astronomical observatories peered to the far depths of the universe.

None of this happened. From 1992 on, under the Russian space program, Buran was canceled, the Mars 96 probe crashed ignominiously, the unmanned scientific program was almost wound up, Russia's military could barely keep watch on its enemies from above and nobody talked about going to Mars any more. Tsiolkovsky's visions of exploring the solar system were relegated to the archive. The historical artefacts of the golden years of Soviet rocketry were pawned off. Aging, retired designers mused nostalgically about the good old days. When cosmonauts did fly into space, now a less frequent event, the two or three helmeted adventurers were accompanied to the launch pad by a man in the black, flowing gowns of a priest of the Orthodox church to bless them on their way. They needed all the luck they could get: quality control was no longer what it was—rockets failed for lack of it. Cosmodromes rotted before their very eyes and the always reliable rocket troops were now mutinous. Staff and technicians left for greener pastures abroad and more enterprising companies at home. The beautiful ships of the tracking fleet had been cut up for scrap. The exhibition hall in Moscow, which had proudly displayed Soviet space achievements and where millions had gasped in wonderment at what had been done, was now a car salesroom. Nobody knew who the cosmonauts were any more and few cared. The space shuttle was a children's amusement in Gorky Park. If ever there was a nemesis of a great and noble project, the Russian space program surely was. In 1996, *Novosti Kosmonautiki* magazine predicted that the program would come to an end in 1998.

Despite all that, the Russian space program clawed its way back. In 2000 Russia regained its place as the top space-faring nation in numbers of rockets launched each year. When the American space shuttle *Columbia* burned up in 2003, it was Russia that kept the International Space Station going, smoothly and without fuss. Against the odds, Russia managed to:

- keep the Mir space station in operation until its safe de-orbiting in 2001;
- build the core modules of the International Space Station, Zarya and Zvezda, as well as supply a docking module, Pirs;
- send a regular supply of Soyuz and Progress missions up to the ISS, including new versions of both: the Soyuz TMA and Progress M1 models;
- maintain a military space program;
- sustain a space applications program.

The Russian space program demonstrated a high level of adaptability to the new, difficult and uncertain economic conditions. This was most clearly demonstrated by:

- the establishment of a national space agency, the RKA, now Roscosmos;
- the turning around of the program from the most self-sufficient national program to the most globally competitive in the world;
- the attraction of significant foreign investment to sustain the manned and un-manned program;
- 87 space-based companies which entered joint ventures with American and European companies to sustain and develop their projects;
- the opening of new cosmodromes (Svobodny and Dombarovska), the develop-ment of new launching systems (Barents Sea) and a launch base in French Guyana;
- the adaptation of missiles to serve as launchers: Rockot, Start, Dnepr and Shtil;
- the introduction of new upper stages: Ikar, Fregat, Briz KM and Briz M.

The Russian space program began to show the promise of new life:

- fresh groups of cosmonauts were recruited;
- the production line of the Soyuz and Proton rockets was increased;
- the Soyuz 2 series was introduced;
- progress was made in the preparation of a new family of rockets, the Angara.

It is possible that 1997 marked the low point of the extreme financial and organiza-tional pressure inflicted on the Russian space program. Ten years later, Russia was in a better position to develop future projects. In 2005 the government approved a new federal space plan. Here we review its key elements, for they mark out the intended future path of Russian space exploration.

Historians of the Soviet space program mourned the closure in the 1990s of the Cosmos pavilion in the Exhibition of Economic Achievements in north Moscow, the VDNK. This was a splendid, if old-fashioned, display area of Soviet space achievements where all the latest space equipment and hardware were put on public show, often for the first time. During the financial shortages, the spaceships were shunted to one side, to be replaced by a car salesroom. In 2006 work began on a new pavilion next door to the old Cosmos pavilion, but one twice as large and intended to be the definitive display area of the Russian space program.

THE FEDERAL SPACE PLAN

In July 2005 the Russian government announced that there would be a new, ten-year civil space plan for 2006–15. The total cost was $20bn of which:

- R130bn ($4.5bn) would be for commercial operations;
- R230bn ($8bn) would be for the federal civilian space program;
- $7.5bn would be for the military space program, including GLONASS.

The key elements were:

- introduction of the Kliper space shuttle and the Parom space tug;
- return to the Moon (Luna Glob);
- return to Mars (Phobos Grunt);
- Mars 500 simulation;
- introduction of the Angara launcher;
- completion of the Russian segment of the International Space Station, with one module launched in the years 2009 (MPLM), 2010 (power station) and 2011 (research module);
- completion of the GLONASS group to 18 operational satellites by end 2007;
- upgrade the program for the R-7 rocket, called Rus-M, leading to the introduction of the Soyuz 3 rocket;
- new Earth resources programs using the Resurs DK platform: Resurs P (2009) and Smotr (2007) and a small platform Arkon (2007);
- earthquake-monitoring satellite Vulkan (2007);
- new science missions: Koronas Foton (2007), Spektr R (Radioastron, 2007), Spektr RG (Radio Gamma, 2009) and Spektr UV (Ultra Violet, 2010), Intergelizond (2011), Venera D (2016), Celsta (2018) and Terion (2018);
- resumption of Bion missions, with Bion M (2010);
- new weather satellites Elektro L (2007) and Electro P (2015).

The federal space plan was formally approved by the government as special resolution 635 of 22nd October 2005. Here we review its key new elements.

REPLACING THE SOYUZ: KLIPER

When Soyuz was designed in 1962 and flown in 1966, no one could have imagined that it would still be flying forty years later. The Russians introduced new versions in the 1970s (Soyuz T), 1980s (Soyuz TM) and 2000s (Soyuz TMA), with the promise of more to come (Soyuz TMS, TMM). Introduction of the TMS was set for 2011. The TMS will:

- carry improved avionics and computers;
- deploy parachutes at a lower altitude, giving greater accuracy with a view to landing in Russian territory, not Kazakhstan;
- extend orbital lifetime to 210 days;
- have two more braking engines, for more precise rendezvous and dockings.

The TMM will have:

- on-orbit lifetime of 380 days through a new preservative system for the hydrogen peroxide fuel;
- a satellite relay system, called Regul;
- a new rendezvous system, KURS MM;
- a further two rendezvous and docking engines [1].

In 2006 Roscosmos began discussion with the European Space Agency (ESA) for European participation and financing for a substantial Soyuz upgrade, which would incorporate many of the ideas of the TMM and TMS. This was later called the CSTS, or Crew Space Transportation System. On the European side, the main participant countries were France, Germany, Italy, Belgium, Spain and Italy, who committed €15m to a two-year study of CSTS. Upgrade options under consideration went as far as looking at a 12-tonne spaceship with a cabin able to fly four cosmonauts. ESA was insistent that its participation in the modernization would be dependent on European companies having a substantial role in designing, building and supplying CSTS components—minor sub-contracting would not be enough [2].

Ultimately, some form of replacement to the Soyuz must be contemplated. Had the Buran space shuttle developed as planned in the 1990s, it might have been possible to phase out Soyuz then. Soyuz was quite small and cramped and, if filled with its normal complement of three cosmonauts, had a spare payload of only 50 kg for experiments and personal effects. Still, Soyuz had considerable cost advantages, costing about €50m a mission. By contrast, the cost of flying the American shuttle was in the order of €500m—the cost savings from reusability proved illusory—and Buran would probably have been similar. During his final days, chief designer Valentin Glushko drew up designs for a large version of the Soyuz, called Zarya, launched by the Zenit and with a crew of six, using lightweight shuttle-type tiles for thermal protection instead of the heavier, old-fashioned heat shield.

News that Russia was working on a complete replacement for the Soyuz came unexpectedly on 17th February 2004 when RKA director Yuri Koptev told a startled

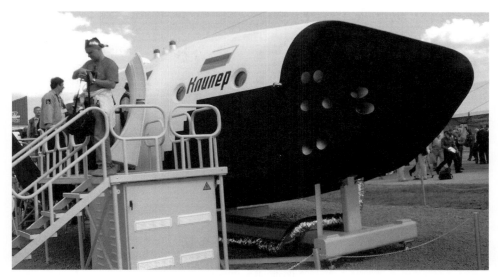

Kliper

press conference, almost in passing, that Russia was working on a new, small space shuttle project. Two days later, RKK Energiya confirmed that it had undertaken studies of such a project on its own initiative since 2000 and that it was called Kliper ("clipper", from the days of the old sailing ships) [3].

Kliper had its origins in studies in the 1990s. During the period of the Mir space station, small cargoes had been returned to Earth in recoverable cone-shaped capsules called Raduga. Their load was less than 150 kg and, with a view to bringing back larger payloads from orbital stations, Energiya studied a set of recoverable maneuverable capsules, called VMKs. These combined lifting bodies with Soyuz components and a version was patented in 1994.

In effect, Kliper represented a development of the VMK studies. Kliper went through many evolutions over the following four years, alternating between winged cabins and lifting body designs. What they had in common was a blunt cabin, in the recognizable shape of a space shuttle type of vehicle and, at the rear, a Soyuz-type docking compartment (although it was called the aggregate compatment). The ideas around Kliper became more apparent when a full-scale model was exhibited at the Moscow MAKS Air Show at Zhukovsky in August 2005. Life-size dummies sat inside, one suited, one in coveralls, so it was possible to get an idea of the scale. Essentially, Kliper was a six-person spaceplane, about 12.5 tonnes in weight, able to spend a year in orbit, a replacement for the Soyuz and intended to be the ideal vehicle for the resupply of the International Space Station in about six years time.

Kliper was intended to reduce payload-to-orbit costs by two-thirds and provide a much smoother journey for the crew, with typical G loads of only 2.5. This in turn would make crew-training less demanding, the medical requirements of cosmonauts less rigorous and reduce crew-training times. Appointed technical director of the

Nikolai Bryukhanov

Kliper's wings

Kliper's nose

project was 1980 graduate of the Moscow Technical University Nikolai Bryukhanov, who had joined RKK Energiya in 1996.

The crew compartment was designed to hold six people, effectively two pilots and four passengers, in a volume of $20\,\mathrm{m}^3$, five times the space of the Soyuz, with two windows on either side. In addition to the human crew, there would be room for 500 kg of cargo. Kliper was designed to ferry crews up to and down from the space station, with a normal independent flying time of five days. One would be attached at all times to the station as a lifeboat. The aggregate compartment, using tested Soyuz components, would provide additional living space, enable docking with the space station (albeit backwards) and would be jettisoned before reentry. The aggregate compartment, with $8\,\mathrm{m}^3$ of space, would provide additional living space, 2,000 kg of fuel for on-orbit maneuvers, the rendezvous and docking system, eight maneuvering engines, eight thrusters and the docking hatch. At the bottom would be eight solid-fuel rocket motors, which would function as a launch escape system, blasting the entire Kliper free of an exploding rocket on the pad. Assuming that this was not required, in a normal mission the solid-fuel rockets would fire to kick the Kliper into orbit, so they would be used anyway.

Kliper would use 60-cm thermal tiles similar to those developed for Buran and the American shuttle, but the nose, which would take the brunt of the reentry, would use the type of material used by Soyuz for its heat shield. Maneuvering would be done by six 24-kg thrusters. Electricity would be provided not by the traditional solar panels but by fuel cells called Foton of the type used on the Buran and able to generate 2.5 kW of electricity. Rudders on the Kliper's wings would enable it to maneuver up to 500 km after reentry, but the lifting body version would not land like the American shuttle or the Buran. Instead, a parachute would open and it would descend vertically. The final touchdown would be cushioned by small rocket engines and shock absorbers. The winged version would, by contrast, make a standard runway-type landing.

Two versions of Kliper were studied by Energiya: a wingless version, which would parachute back to landing with a landing accuracy of 15 km; and a more streamlined winged version, able to land on a runway. The wingless version was cheaper to build and operate, but the winged version offered better performance and a smoother landing. During 2006, thermal tests were carried out in the Sukhoi Design Bureau of the best shape for the winged version, using materials developed during the building of the shuttle Buran. The outcomes were encouraging, for with computer-aided design it was possible to achieve temperatures 500° better than Buran. Almost all the thermal impact was on the nose. Kliper's wings were further back than the American space shuttle, reducing the danger of the type of problem that destroyed *Columbia*. Interestingly, when the European Space Agency engineers came to look at future possible European spaceplane designs later that year, they came up with a shape almost identical to Kliper.

Just as the Kliper design went through many evolutions, so too did the launcher intended for it. Initially, the new Onega launcher was considered to be the prime candidate to launch Kliper, but outgoing Ukrainian president Leonid Kuchma proposed the Ukrainian-built Zenit and this was accepted by President Putin. Zenit had the advantage of being a proven rocket, planned for manned spaceflight, but with the disadvantage that there was only one Zenit launch pad, the second one having been wrecked and not rebuilt following an explosion in 1990. The election of Viktor Yushenko to the Ukrainian presidency the following month appears to have cast doubt on the Zenit idea, and it was learned later that still more derivatives of the Soyuz were under consideration as Kliper's launcher. As of 2006, a version of the Soyuz was still the favored launcher, with Zenit still a possibility.

The Kliper remained a paper project until 2005, developed by Energiya's technical staff at the company's own expense. Prospects for the Kliper improved that summer, when it was specifically included in the federal space program for 2006–15, with a budget of R10bn (€300m), with provision for a first flight in 2013 and twenty by the end of the plan. The Russian Space Agency announced that the design would be put out to tender, with three expected contenders: Energiya, Khrunichev and NPO Molniya, the last being a company with experience in spaceplane design. Most people saw this as a formality, for the tender reflected what Energiya had already designed and suspected that this was done to achieve the appearance of open market competition.

At this stage the plot thickened. Funding remained a problem at the heart of Kliper's development and later that year Russia made a strong pitch to interest the European Space Agency. The Europeans baulked, seeing no reason to invest money

Kliper on Onega launcher

Kliper docking system

without the prospect of seeing some of the development or construction work come their way or of obtaining some other defined advantages. In June 2006 the European Space Agency did agree to invest €15m in the joint Russian–European study the Crew Space Transportation System (CSTS), which included the Soyuz upgrade described above and options beyond. The tender was promptly suspended while this took place. The Russian intention was that the joint study would confirm the wisdom of the Kliper proposal and, even if it meant a delay of two years, would buy the Europeans into the system.

Soon after the first Kliper designs appeared, an accompanying project was announced, called Parom. The intention of Parom was that it would be an unmanned design that would replace the venerable Progress freighter, which had been resupplying orbital stations since 1978. Parom was a combined space freighter and tug: cargoes to the space station would be launched into a low-Earth orbit in a container called Parom Kh, able to carry 7.6 tonnes of freight, three times that of Progress (2.4 tonnes). Already in orbit would be a tug called Parom B (12,500 kg when fueled, 5,990 kg when dry). This would be moored to and based at the International Space Station for fifteen years, but as soon as a Parom Kh was launched, the Parom B would descend, rendezvous, dock and tug the Parom Kh cargo up to the station. Between them, they would treble the amount of freight that could be brought up to the station at a time. Originally, Parom was to follow Kliper, but in October 2006 the order was reversed and the first test flight for Parom was set for December 2010. It was a neat idea and Energiya claimed it could achieve considerable cost savings over Progress as well as the European and Japanese freighter systems for the International Space Station.

Some of Russia's space future depends on the later stages of the building of the International Space Station. The leading role of the United States in the station undoubtedly alienated some in the Russian space community, who proposed that Russia build its own, national space station flying at 65° to the equator and able to survey the country's northern landmass. Proposals toward this effect appeared from time to time. Once the Americans embarked on the Visions for Space Exploration program, it became evident that they would lose interest in the International Space Station, giving Russia a free hand to resume the leading role. Proposals for a high-latitude space station gradually disappeared from the menu.

RETURN TO THE MOON: LUNA GLOB

Included in the new federal space plan to 2015 was a return to the Moon in 2012 [4]. In the post-Soviet space program, the Moon had been rarely mentioned. The only time it was discussed was in summer 1997 when the Institute for Space Research, IKI, proposed plans to send a small spacecraft into lunar orbit, using a Molniya rocket from Plesetsk Cosmodrome in northern Russia in 2000. The orbiter would deploy three 250-kg penetrators, to dive into the lunar surface at some speed, burrowing seismic and heat flow instruments under the lunar surface, leaving transmitters just above the surface. With small nuclear isotopes, they would transmit for a year, operating as a three-point network to collect information on moonquakes and heat flow. A number of variations on this theme appeared, but none progressed beyond the aspirational stage. Over time, this mission acquired the title Luna Glob, or "lunar globe", presumably from the global nature of the seismometer system.

Details of a new Luna Glob mission were given by officials of the Russian Space Agency, the Vernadsky Institute and the Institute of Earth Physics in 2006. All appeared anxious that Russia, for all its past expertise in the exploration of the Moon, should get back in the business of lunar exploration. They were also motivated not just by American plans to return to the Moon by 2020 announced by President Bush, but by the prospect of Moon probes being sent there by China, India and Japan much sooner.

The new Luna Glob envisaged the launch by a Molniya rocket of a 1,500-kg mother ship into lunar orbit. Before arriving at the Moon, the mother ship would release a fleet of ten high-speed 30-kg penetrators to impact into the Sea of Fertility in a circular pattern, each only 2,500 m from the next one, forming a ten-point seismic station. The mother ship would continue into lunar orbit. Next, it would deploy two penetrator landers at the Apollo 11 and 12 landing sites, to rebuild the seismic network the Americans began there in 1969. Then, it would send a soft lander down to the south polar region, called the polar station, carrying a seismometer and two spectrometers to detect water ice. The mother ship would act as a relay for the thirteen data stations on the lunar surface (a small moonrover of 150 kg was also in consideration). If Luna Glob were successful, then a 700-kg rover might follow in 2015–16. The Lavochkin Design Bureau also sketched a sample return mission to follow, the design looking like a smaller, neater model of the Luna sample return missions of the 1970s. The space agency emphasized that Luna Glob was a real, planned mission—not a paper project.

The Moon was the first target beyond the Earth for the old Soviet space program, the first probes being sent there in 1959. Venus was the second, in 1961, a planet where the Soviet Union achieved remarkable successes, but which was last visited in 1985. The federal space plan included a return to Venus, with a Venera mission sketched for some time between 2012 and 2016. With a Soyuz 2 Fregat launcher, the spacecraft would weigh 1,400 kg. Two missions were under consideration. The first was a mission that had originally been planned in the early 1980s: a long-duration Venus lander, to survive on the boiling hot surface for up to a month with the objective of detecting seismic activity. The second possibility was a mixed mission probe combining a 30-kg small lander, a 100-kg atmospheric probe and a 100-kg balloon.

RETURN TO MARS: PHOBOS GRUNT

During the 1970s the Lavochkin Design Bureau had made a number of studies of the possibility of recovering samples from Mars. At one stage, this was given the go-ahead in the form of the hugely ambitious and complex project 5M between 1975 and 1977. Although the manufacturing of hardware began, the project was canceled.

But, if collecting samples from Mars was too difficult, what about Mars' small moon Phobos? Russia had a good knowledge of Phobos, for the spacecraft Phobos 2 had closed in on the small moon in March 1989, although communications were lost before it could send down a lander. The Institute for Space Research, IKI had floated the notion of converting the flight model of Phobos 3 for such a mission. Instead, Russia's only Mars probe in the 1990s was a large orbiter and landing mission, Mars 96. Postponed and repeatedly run on an on–off basis, nothing symbolized its diffi-culties more than the fact that it was fitted out in the Baikonour cosmodrome by candlelight because there was no money to pay for electricity. The probe was lost within a day of launch.

In 1999 the Russian Academy of Sciences, through the Institute of Space Research (IKI) and OKB Lavochkin began a feasibility study of recovering rock samples from the moon Phobos and invested an initial R9m in the project. Called Phobos Grunt (literally in Russian "Phobos Soil"), the probe was similar to, but smaller than, the failed Phobos missions of 1988–89 and would follow a similar mission profile. It would use the 8K78M Molniya launcher (the Proton was too expensive) and electrical engines to get the spacecraft out of Earth orbit on its way to Mars.

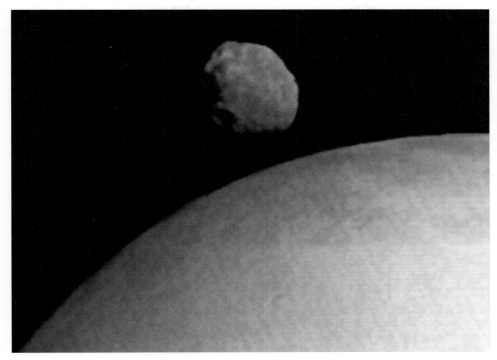

Phobos, target of Phobos Grunt

The 1999 study was re-worked in 2003, with a revised specification. The launcher proposed was the now-proven Soyuz Fregat. The electrical propulsion system, called the SPT-140, was upgraded to three thrusters using 4.5 kW power drawn from very large solar wings. The mission specification was for Soyuz Fregat to put the Phobos Grunt in a high-Earth orbit of of $215 \times 9{,}385$ km. At this stage, the Fregat stage would be dropped and the solar electric power take over to spiral the spacecraft, now 4,660 kg, out of Earth orbit and toward Mars, the huge solar panels being dropped at Mars orbital insertion. Once in Martian orbit, the spacecraft would follow the path of Phobos 2 in 1989, entering an orbit that passed Phobos every four days. Small rockets would be used to land and then press the spacecraft onto the surface and anchor it in the low gravity. The entire spacecraft was designed to land and remain on Phobos for some time, powered by solar panels. A drill would be used to collect soil samples, modeled on the system used to collect Moon rock in the 1970s. A couple of days after scooping up about 170 g of rock samples from Phobos, a small capsule with the samples would be fired toward Earth, to be recovered 280 days later in Russia. The Russians believed that they could build a spacecraft with a surface payload of 120 kg, which is substantial and attractive from the point of view of international participation [5]. From 2004 the Russians began to sound out, informally, the prospects of such international cooperation. Already, a prospective instrument manifest was drafted.

Prospective instruments, Phobos Grunt
TV system
Gamma ray spectrometer
Neutron spectrometer
Alpha X spectrometer
Mass spectrometer
Seismometer
Long-wave radar
Visual and near-infrared spectrometer
Dust counter
Ion spectrometer
Solar sensor

For many years, Phobos Grunt looked like another paper project that was never going anywhere. It was a repeated theme on the wish list of Soviet planetary scientists and engineers. Their fixity of purpose eventually paid off, for when the ten-year Soviet space plan for 2006–15 was published in summer 2005, Phobos Grunt was one of the new flagship projects included. Launch was set for October 2009, reaching the red planet a year later, leaving Phobos in summer 2011 and returning to Earth in June 2012. A project manager, Igor Goroshkov, was appointed. Detailed sketches of the revised Lavochkin design were published. There were further revisions to the 2003 plan, though the flight profile was much the same. The two-tonne spacecraft was now a low, squat, neat polygon on four legs, with a new type of drilling rig at the side, with a long tube for vacuuming the soil sample into the recovery cabin. The lander would work on Phobos for a year after the return stage was fired back to Earth, equipped with about 20 science experiments provided by Russia and the international science community. The return stage was a small, box-shaped module with a solar panel on one side, the ball-shaped sample recovery cabin fitting snugly in the middle. The return stage would lift off the lander, enter Martian orbit, orientate itself, fire to Earth, transit for eleven months and then release the small sample recovery cabin for its plunge into the atmosphere. The small size of the spacecraft, benefiting from recent advances in electronics, miniaturization and materials, was a dramatic contrast to the size and complexity of Sergei Kryukov's project in the second half of the 1970s.

Later, if all this went well, there was the prospect of a return to the Martian surface, with a small six-wheel rover (95 kg) and then a sample return mission (Mars Grunt). Designs were drawn up by the Lavochkin Design Bureau in 2006, showing how a dome-shaped lander would enter the Martian atmosphere protected within an inflatable rubber braking cone and fire rockets for the final stage. Once a robotic arm had selected and retrieved samples, a small rocket in the top of the dome would blast earthward. Lavochkin ambitiously targeted the mission for 2012, but much would depend on the outcome of Phobos Grunt.

Mars sample return

MARS 500: NO GIRLS PLEASE, WE'RE GOING TO MARS

Granted the acute difficulties in getting an unmanned mission to Mars under way at all, the idea of even considering manned missions to the planet might seem surreal. Yet, that is exactly what happened. The Russians saw no reason not to work away at the critical paths necessary for a long-duration mission which could take place in better times. Three previous simulated Martian missions had been carried out in modules called *bochkas* in the Institute for Medical and Biological Problems (IMBP) (365 days in 1967, a 370-day mission in 1986 and the 240-day controversial and conflictual Sphinx mission, 2000). Now, IMBP's Mark Belkovsky announced that, with the backing of both the Russian Space Agency and the Russian Academy of Sciences, Russia would recruit 20 volunteers for a simulated 520-day flight to Mars. Twenty volunteers were sought: two groups of six as model "crews" and a control group, with international participation invited. Funding had been put together from three sources: the federal space budget (R160m), the Academy of Sciences and the Ministry of Education and Science. A technical director was appointed, Yevgeni Demin.

He received several immediate enquiries about participating in the experiment, with two real cosmonauts indicating their interest. The experiment would mimic flight time outbound (250 days), Mars landing (30 days) and return (240 days) and the crew would be expected to follow a regime of experiments, exercise, maintenance and even simulated emergencies. The crew would have three tonnes of water and five tonnes of food, generating oxygen by a closed-cycle life support system. The experiment would be carried out in parts of the original *bochka* used for the 1967 experiment, but with the addition of two mock Mars surface modules, bringing the volume of the four-module living space up to 500 m^3. By early 2007, construction of the new modules was well under way. To make the simulation more realistic, there would be longer and longer delays in the time ground control would take to respond to messages from the *bochka*—40 min by the time they reached Mars itself. Systems would be put in place to study all the key factors of such a mission: climate, immune systems, toxicology, plant nursery, psychology. The experiment would operate a Martian day of 24 hr 40 min.

In a controversial move, the IMBP announced that only men would be selected, a point elaborated by the IMBP director Anatoli Grigoriev, who described a mission to Mars as "too demanding" for women. Equal rights campaigners abroad at once demanded their countries not participate unless this condition be lifted. One of the participants would be a doctor and his work would be an important test of tele-medicine systems—or as one of the project directors said, "if they have a fight this time, they'll have to sort it out themselves" [6]. By 2007, 150 volunteers had stepped forward for the mission, including—notwithstanding Gregoriev's view—16 women and one married couple. Apart from Russia, they came from the United States, Spain, Ukraine, India and Australia. The European Space Agency volunteered to cover the costs of two Europeans to participate in the experiment. The first five volunteers were selected in January 2007.

More mundanely, but nonetheless important, was on-going Russian work on closed-cycle environments. Such work had started in the 1960s in what were called

The *bochka*

biospheres (bios for short) in the Physics Institute of the Siberian Department of the Academy of Sciences in Krasnoyarsk, where scientists built sealed greenhouses where plants generated oxygen and produced food for future Mars or lunar colonists. The main food grown was salad, parsley and carrots. The center was able to attract foreign foundation grants after 2000, maintaining its record as the world leader in the field. The initial concentration of the biospheres on vegetables may pay off in the end, for experiments on the Salyut and Mir space stations with growing quail and hatching fish proved unsuccessful. The chicks could not adapt to zero gravity, while fish grew far too slowly, leaving cosmonauts with the prospects of neither fish nor fowl but vegetables instead.

Not only was work going on at the medical end, but some of the key technologies for a Mars mission were studied. As far back as the 1950s, the predecessor to the Energiya Design Bureau, OKB-1, had designed a prototypical manned spaceship to fly to Mars, the TMK, and had revised these designs many times over in subsequent years: in the 1960s (Aelita), 1970s and 1980s. In 1999, Energiya re-iterated its Mars concept designs, making electrical propulsion a key feature of its sketch.

An international science and technology committee was set up in 2001 comprising representatives of Russia (eight), the United States (eight) and the European Union (five) to facilitate coordination between national space programs, in general, and Mars exploration, in particular. On the Russian side, the project involved the federal scientific centers, RKK Energiya, the Institute for Space Research (IKI) and the Institute for Medical and Biological Problems (IMBP). RKK Energiya quickly contributed to the committee the design of a manned Martian orbital station called MARSPOST (MARS Piloted Orbital STation), which would serve as a base for cosmonauts to drop probes from Mars orbit. MARSPOST was the idea of Leonid Gorshkov of RKK Energiya who the previous year had combined some of the ideas of the old TMK-1 design for a Mars flyby with the accumulated experience of orbital stations since then. Meantime, the Keldysh Research Center had already tested a 1:20-scale Martian descent module, designed to operate with MARSPOST. The descent module would bring cosmonauts down to the planet, carry a rover for its exploration and enable them to return to the MARSPOST afterwards. Energiya sketched out a 730-day mission for a crew of six. MARSPOST would have a mission commander, flight engineer and doctor. The surface expedition would have a pilot, biologist and geologist to spend 30 days there.

Energiya refreshed its Mars designs in 2005. The key elements of its 2005 plan were:

- an interplanetary crew module, shaped like the Mir base block, with a crew of six in six pressurized sections of $410 \, m^3$ volume;
- large solar electric propulsion array of 30 kW;
- electric engines, used for acceleration out of Earth's gravity (three months), outbound (eight months), deceleration into Mars orbit (one month), acceleration from Mars orbit (one month), earthbound (seven months) and deceleration into Earth's gravitational field (one month);
- Mars lander including as its top stage an ascent module. The lander would be 62

tonnes in mass, 40 tonnes on the Mars surface, with an ascent module mass of 22 tonnes, with a capsule of mass 4.3 tonnes.

Energiya repeatedly reminded Russia's leaders that it had done all the necessary homework for such a mission. All it needed was the go-ahead—and the money. Ultimately, of course, the Russians would like to resume progress on their great schemes for Moon and Mars bases envisaged during the heady days of the Soviet period. The federal space plan (Table 8.1) envisages the taking of such concepts a step further in the federal plan, to begin 2016, and had pencilled in the construction of

Table 8.1. Year-by-year, the Russian federal space plan 2006–15

Year	Mission name	Mission
2007	Koronas Foton	Solar science
	Spektr R	Radio astronomy
	Smotr	Gasfields
	Arkon	Earth resources
	Vulkan	Earthquake detection
	Elektro L	24-hr weather satellite
	Yamal GK	Communications
	Ekspress AM	Communications
	Meteor 3M2	Weather satellite
2008	Luch 5	Data relay for ISS
	Ekspress V	Communications
	Yamal 200M	Communications
2009	MPLM	Space station
	Phobos Grunt	Phobos sample return
	Spektr RGRL	X-ray astronomy
	Resurs P	Earth resources
2010	Power module	Space station
	Parom	New freighter
	Bion M	Biology
	Ekspress AT	Communications
	Yamal 300	Communications
	Spektr UV	Ultra-violet astronomy
2011	Research module	Space station
	Soyuz TMS	New Soyuz version
	Intergelizond	Solar probe
2012	Meteor PM	Weather satellite
	Luna Glob	Moon probe
2015	Elektro P	24-hr weather satellite
2016	Venera D	Venus mission

Moon and Mars bases in the 2025–50 period. Soviet experts had designed moonbase Galaktika in 1969 and a second one, Zvezda, in 1974. In 2006, sketches appeared of what was called Base 2050, fully kitted out with laboratories, fuel-processing stations, rovers, scientific stations, an underground adaptation and rehabilitation center and even a conference hall.

FINAL REMARKS

The commitment of the Russians to their space program was something which many outside observers found hard to understand. It is not one universally shared in a country which has endured much hardship and where many people have more immediate and pressing concerns on their mind, but it is one held by enough people to matter. Soviet and Russian achievements in space were built up painstakingly, painfully, over many years. Their founders had learned in the hard school of the camps, the wartime frontline, the early rockets that often exploded and the Brezhnevite bureaucracy. They had known the heartbreaking failures, the loss of two Soyuz crews, the satellites that went silent, the upper stages that would not be tamed, the Moon race they could not win, the Mars probes that disappeared. But, they also remembered the night the Sputnik was launched, the day they hit the Moon, the glory of Gagarin's flight, Tereshkova, the spacewalk, the soft landing on the Moon, the pictures from Venus, the first space station and then Mir. These things had enabled the Soviet Union and Russia to walk tall in the world, to mark out space exploration as a unique arena of accomplishment. It was a space program in which its participants and admirers could justifiably take immense pride, a program built on a potent mixture of courage, endurance, daring, engineering genius, quality and imagination. It was a program which had deep historical roots—going back to Tsiolkovsky in the 1890s, Kondratyuk's writings during the First World War, Tsander's plans to go to Mars, Glushko's first experiments in the Gas Dynamics Laboratory in the 1920s. It was a program which both pre-dated and outlived the communist experiment. In keeping the Russian space program going, its engineers and scientists, now joined by its managers and accountants, were keeping alive a dream that went back two centuries.

Television viewers watching astronauts and cosmonauts gathering in the control cabin of the International Space Station for their group picture may puzzle who is the monocled, bearded old man whose picture appears in the background, but it's the old man, the dreamer, Tsiolkovsky. He himself said it best, in 1897:

> Mankind will not remain forever on the Earth. In pursuit of light and space he will timidly at first probe the limits of the atmosphere and later extend his control throughout the solar system. Man will ascend into the expanse of the heavens and found settlements there. The impossible of today will become the possible of tomorrow.

REFERENCES

[1] Hall, Rex D. and Shayler, David J.: *Soyuz—a universal spacecraft*. Springer/Praxis, 2003.

[2] Frédéric Castel: Cooperation on a new Soyuz. *Planet Aerospace*, 1/2007.

[3] Hendrickx, Bart: In the footsteps of Soyuz—Russia's Kliper spacecraft, from Brian Harvey (ed.): *2007 Space exploration annual*, Springer/Praxis, 2006.

[4] Covault, Craig: Russia's lunar return. *Aviation Week and Space Technology*, 5th June 2006; Craig Covault: Russian exploration—Phobos sample return readied as Putin's government weighs Moon/Mars goals. *Aviation Week and Space Technology*, 17th July 2006

[5] Popov, G.A.; Obukhov, V.A.; Kulikov, S.D.; Goroshkov, I.N. and Upensky, G.R.: *State of the art for the Phobos Soil return mission*. Paper presented to 54th International Astronautical Congress, Bremen, Germany, 29th September–3rd October 2003; Ball, Andrew: *Phobos Grunt—an update*. Paper presented to the British Interplanetary Society, 5th June 2004.

[6] Parfitt, Tom: Spaceflight is hell on Earth. *The Guardian*, 8th September 2005. See also: Oberg, Jim: Are women up to the job of exploring Mars? *MSNBC*, 11th February 2005; Phelan, Dominic: Russian space medicine still aims for Mars. *Spaceflight*, vol. 46, #1, January 2004; Zaitsev, Yuri: Preparing for Mars—a simulated manned mission to the red planet is about to begin. *Spaceflight*, vol. 47, no. 1, January 2005.

Appendix

Launchings 2000–6

This listing is defined as launchings in which satellites reached orbit. For a list of launchings 1992–2000, see the previous edition, *Russia in space – the failed frontier?*

2000

1 Feb	Progress M1-1	Soyuz U	Baikonour	
3 Feb	Cosmos 2369	Zenit 2	Baikonour	Tselina 2
9 Feb	Dumsat/IRDT 1	Soyuz Fregat	Baikonour	
12 Feb	Garuda	Proton K block D	Baikonour	
12 Mar	Ekspress A-2	Proton K block D	Baikonour	
20 Mar	Dumsat	Soyuz Fregat	Baikonour	
6 Apr	Soyuz TM-30	Soyuz U	Baikonour	
18 Apr	Sesat	Proton K block D	Baikonour	
26 Apr	Progress M1-2	Soyuz U	Baikonour	
3 May	Cosmos 2370	Soyuz U	Baikonour	Yantar Neman
16 May	Simsat 1, 2	Rockot	Plesetsk	
6 Jun	Gorizont 33	Proton K Briz M	Baikonour	
24 Jun	Ekspress A-3	Proton K block D	Baikonour	
28 Jun	Nadezhda M-6 Tsinghua SNAP 1	Cosmos 3M	Plesetsk	
30 Jun	Sirius 1	Proton K block D	Baikonour	
5 Jul	Cosmos 2371	Proton K block D	Baikonour	Potok

2000 (*cont.*)

12 Jul	Zvezda	Proton K	Baikonour	
15 Jul	Champ Meta Rubin 1	Cosmos 3M	Plesetsk	
16 Jul	Cluster 1	Soyuz Fregat	Baikonour	
28 Jul	PAS 9	Zenit 3SL	*Odyssey* platform	
6 Aug	Progress M1-3	Soyuz U	Baikonour	
9 Aug	Cluster 2	Soyuz Fregat	Baikonour	
28 Aug	Raduga 1-5 Globus 1	Proton K block D	Baikonour	
5 Sep	Sirius 2	Proton K block D	Baikonour	
25 Sep	Cosmos 2372	Zenit 2	Baikonour	Orlets Yenisey
27 Sep	Megsat Unisat Saudisat 1A Saudisat 1B Tiungsat	Dnepr	Baikonour	
29 Sep	Cosmos 2373	Soyuz U	Baikonour	Kometa
2 Oct	GE-1A	Proton K block D	Baikonour	
13 Oct	Cosmos 2374-6	Proton K block D	Baikonour	GLONASS
17 Oct	Progress M-43	Soyuz U	Baikonour	
21 Oct	Thuraya	Zenit 3SL	*Odyssey* platform	
22 Oct	GE-6A	Proton K block D	Baikonour	
31 Oct	Soyuz TM-31	Soyuz U	Baikonour	
16 Nov	Progress M1-4	Soyuz U	Baikonour	
30 Nov	Sirius 3	Proton K block D	Baikonour	
5 Dec	Eros A	START 1	Svobodny	

2001

24 Jan	Progress M1-5	Soyuz U	Baikonour
20 Feb	Odin	START 1	Svobodny
26 Feb	Progress M-44	Soyuz U	Baikonour
18 Mar	XM Rock	Zenit 3SL	*Odyssey* platform
7 Apr	Ekran M4	Proton M	Baikonour
28 Apr	Soyuz TM-32	Soyuz U	Baikonour

7 May	XM Roll	Zenit 3SL	*Odyssey* platform	
13 May	Panamsat 10	Proton K block D	Baikonour	
20 May	Progress M1-6	Soyuz FG	Baikonour	
29 May	Cosmos 2377	Soyuz U	Plesetsk	Kobalt
8 Jun	Cosmos 2378	Cosmos 3M	Plesetsk	Parus
16 Jun	Astra 2C	Proton K block D	Baikonour	
20 Jul	Molniya 3-51	Molniya M	Plesetsk	
31 Jul	Koronas F	Tsyklon 3	Plesetsk	
21 Aug	Progress M-45	Soyuz U	Baikonour	
24 Aug	Cosmos 2379	Proton K block D	Baikonour	Prognoz
15 Sep	Pirs	Soyuz U	Baikonour	
6 Oct	Raduga 1-6	Proton K block D	Baikonour	
21 Oct	Soyuz TM-33	Soyuz U	Baikonour	
25 Oct	Molniya 3-52, 3K1	Molniya M	Plesetsk	
25 Nov	Progress M1-7	Soyuz FG	Baikonour	
1 Dec	Cosmos 2380-2	Proton K block D	Baikonour	GLONASS
10 Dec	Meteor 3M1 KOMPASS 1 Badr Tubsat Maroc Reflektor	Zenit 2	Baikonour	
21 Dec	Cosmos 2383	Tsyklon 2	Baikonour	US P
28 Dec	Cosmos 2384-6 Gonetz D1 1,2,3	Tsyklon 3	Plesetsk	Strela 3

2002

25 Feb	Cosmos 2387	Soyuz U	Plesetsk	Kobalt
17 Mar	GRACE 1, 2	Rockot	Plesetsk	
21 Mar	Progress M1-8	Soyuz FG	Baikonour	
30 Mar	Intelsat 903	Proton K block D	Baikonour	
1 Apr	Cosmos 2388	Molniya M	Plesetsk	Oko
25 Apr	Soyuz TM-34	Soyuz U	Baikonour	
8 May	Tempo 1/Direct TV 5	Proton K block D	Baikonour	
29 May	Cosmos 2389	Cosmos 3M	Plesetsk	Parus

2002 (*cont.*)

10 Jun	Ekspress A-4	Proton K block D	Baikonour	
15 Jun	Galaxy 3C	Zenit 3SL	*Odyssey* platform	
19 Jun	Iridium 97, 98	Rockot	Plesetsk	
16 Jun	Progress M-46	Soyuz U	Baikonour	
18 Jul	Cosmos 2390-1	Cosmos 3M	Plesetsk	Strela 3?
12 Jul	IRDT 2	Volna	Barents Sea	
25 Jul	Cosmos 2392	Proton K block D	Baikonour	Araks 2
22 Aug	Echostar 8	Proton K block D	Baikonour	
25 Sep	Progress M1-9	Soyuz FG	Baikonour	
26 Sep	Nadezhda M	Cosmos 3M	Plesetsk	
17 Oct	Integral	Proton K block D	Baikonour	
30 Oct	Soyuz TMA-1	Soyuz FG	Baikonour	
(25 Nov	Astra 1K	Proton K	Baikonour)	
28 Nov	Mozhayets 3 Alsat 1 Rubin 3	Cosmos 3M	Plesetsk	
20 Dec	Unisat 2 Saudisat 1C Rubin 2 Latinsat 1 Latinsat 2 Trailblazer	Dnepr	Baikonour	
24 Dec	Cosmos 2393	Molniya M	Plesetsk	Oko
25 Dec	Cosmos 2394-6	Proton K block D	Baikonour	GLONASS
29 Dec	Nimiq 2	Proton M	Baikonour	

2003

2 Feb	Progress M-47	Soyuz U	Baikonour	
2 Apr	Molniya 1-92, 1T-28	Molniya M	Plesetsk	
24 Apr	Cosmos 2397	Proton K block D	Baikonour	Prognoz
26 Apr	Soyuz TMA-2	Soyuz FG	Baikonour	
2 Jun	Mars Express	Soyuz FG Fregat	Baikonour	
4 Jun	Cosmos 2398	Cosmos 3M	Plesetsk	Parus
7 Jun	AMC-9/GE-12	Proton K Briz M	Baikonour	

8 Jun	Progress M1-10	Soyuz U	Baikonour	
10 Jun	Thuraya 2	Zenit 3SL	*Odyssey* platform	
19 Jun	Molniya 3-53	Molniya M	Plesetsk	
30 Jun	Monitor E model Mimosa Most Cubesat CUTE-1 CanX-1 AAU Cubesat DTUsat Quakesat	Rockot	Plesetsk	
8 Aug	Echostar 9	Zenit 3SL	*Odyssey* platform	
12 Aug	Cosmos 2399	Soyuz U	Baikonour	Don
19 Aug	Cosmos 2400-1	Cosmos 3M	Plesetsk Strela 3	
30 Aug	Progress M-48	Soyuz U	Baikonour	
27 Sep	Mozhayets 4 Laretz UK-DMC Nigeriasat Bilsat KASat Rubin 4	Cosmos 3M	Plesetsk	
30 Sep	Galaxy 13/Horizons 1	Zenit 3SL	*Odyssey* platform	
18 Oct	Soyuz TMA-3	Soyuz FG	Baikonour	
30 Oct	SERVIS 1	Rockot	Baikonour	
24 Nov	Yamal 100a, b	Proton K block D	Baikonour	
5 Dec	Kondor model	Strela	Baikonour	
10 Dec	Cosmos 2402-4	Proton K Briz M	Baikonour	GLONASS
28 Dec	Amos 2	Soyuz Fregat	Baikonour	
29 Dec	Ekspress AM-22	Proton K block D	Baikonour	

2004

10 Jan	Telstar/Estrela del Sul	Zenit 3SL	*Odyssey* platform
29 Jan	Progress M1-11	Soyuz U	Baikonour
18 Feb	Molniya 1-93, 1T-29	Molniya M	Plesetsk
16 Mar	Eutelsat W3A	Proton M	Baikonour
27 Mar	Raduga 1-7	Proton K block D	Baikonour

2004 (*cont.*)

19 Apr	Soyuz TMA-4	Soyuz FG	Baikonour	
27 Apr	Ekspress AM-11	Proton K block D	Baikonour	
4 May	Direct TV 7S	Zenit 3SL	*Odyssey* platform	
25 May	Progress M-49	Soyuz U	Baikonour	
28 May	Cosmos 2405	Tsyklon 2	Baikonour	US PM
10 Jun	Cosmos 2406	Zenit 2	Baikonour	Tselina 2
16 Jun	Intelsat 10	Proton M	Baikonour	
29 Jun	Apstar 5/Telstar 18	Zenit 3SL	*Odyssey* platform	
29 Jun	Demeter Unisat 3 Saudisat 2 Saudicomsat 1 Saudicomsat 2 Latinsat C Latinsat D Amsat Echo	Dnepr	Baikonour	
22 Jul	Cosmos 2407	Cosmos 3M	Plesetsk	Parus
4 Aug	Amazonas	Proton M	Baikonour	
11 Aug	Progress M-50	Soyuz U	Baikonour	
23 Sep	Cosmos 2408-9	Cosmos 3M	Plesetsk	Strela 3
24 Sep	Cosmos 2410	Soyuz U	Plesetsk	Kobalt
14 Oct	Soyuz TMA-5	Soyuz FG	Baikonour	
15 Oct	AMC-15	Proton M	Baikonour	
30 Oct	Ekspress AM-1	Proton K block D	Baikonour	
9 Nov	Oblik	Soyuz 2-1a	Plesetsk	
23 Dec	Progress M-51	Soyuz U	Baikonour	
24 Dec	Sich 1M KS5MF2	Tsyklon 3	Plesetsk	
25 Dec	Cosmos 2411-3	Proton K block D	Baikonour	GLONASS

2005

20 Jan	Cosmos 2414 Tatyana	Cosmos 3M	Plesetsk	Parus
3 Feb	AMC-12	Proton M	Baikonour	
28 Feb	Progress M-52	Soyuz U	Baikonour	

1 Mar	XM-3 Rhythm	Zenit 3SL	*Odyssey* platform	
30 Mar	Ekspress AM-2	Proton K block D	Baikonour	
15 Apr	Soyuz TMA-6	Soyuz FG	Baikonour	
26 Apr	Spaceway 1	Zenit 3SL	*Odyssey* platform	
22 May	DirecTV-8	Proton M	Baikonour	
31 May	Foton M-2	Soyuz U	Baikonour	
16 Jun	Progress M-53	Soyuz U	Baikonour	
24 Jun	Telstar 8	Zenit 3SL	*Odyssey* platform	
24 Jun	Ekspress AM-3	Proton K block D	Baikonour	
13 Aug	Galaxy 14	Soyuz FG Fregat	Baikonour	
24 Aug	OICETS INDEX	Dnepr	Baikonour	
26 Aug	Monitor E	Rockot	Plesetsk	
2 Sep	Cosmos 2415	Soyuz U	Baikonour	Kometa
8 Sep	Progress M-54	Soyuz U	Baikonour	
9 Sep	F1R	Proton M	Baikonour	
1 Oct	Soyuz TMA-7	Soyuz FG	Baikonour	
27 Oct	Mozhayets 5 Sinah 1 China DMC SSET Express XIVI UWE NCube 2 Topsat Rubin 5	Cosmos 3M	Plesetsk	
8 Nov	Inmarsat 4	Zenit 3SL	*Odyssey* platform	
9 Nov	Venus Express	Soyuz FG Fregat	Baikonour	
21 Dec	Progress M-55	Soyuz U	Baikonour	
21 Dec	Gonetz D1M Cosmos 2416	Cosmos 3M	Plesetsk	Strela 3
25 Dec	Cosmos 2417-9	Proton K block D	Baikonour	GLONASS
28 Dec	Giove 1	Soyuz Fregat	Baikonour	
29 Dec	AMC-23	Proton M	Baikonour	

2006

15 Feb	Echostar 10	Zenit 3SL	*Odyssey* platform	
(1 Mar	Arabsat 4A	Proton M	Baikonour)	
30 Mar	Soyuz TMA-8	Soyuz FG	Baikonour	
12 Apr	JCSat 9	Zenit 3SL	*Odyssey* platform	
24 Apr	Progress M-56	Soyuz U	Baikonour	
25 Apr	Eros B	START	Svobodny	
3 May	Cosmos 2420	Soyuz U	Plesetsk	Kobalt M
26 May	KOMPASS 2	Shtil	Barents Sea *Ekaterinberg*	
15 Jun	Resurs DK	Soyuz U	Baikonour	
18 Jun	Galaxy 16	Zenit 3SL	*Odyssey* platform	
19 Jun	Kazsat	Proton K block D	Baikonour	
25 Jun	Cosmos 2421	Tsyklon 2	Baikonour	US P
28 Jun	Progress M-57	Soyuz U	Baikonour	
12 Jul	Genesis 1	Dnepr	Dombarovska	
21 Jul	Cosmos 2422	Molniya M	Plesetsk	Oko
28 Jul	Kompsat 2	Rockot	Plesetsk	
4 Aug	Hot Bird 8	Proton M	Baikonour	
14 Sep	Cosmos 2423	Soyuz U	Baikonour	Don
22 Aug	Koreasat 5	Zenit 3SL	*Odyssey* platform	
18 Sep	Soyuz TMA-9	Soyuz FG	Baikonour	
19 Oct	Metop A	Soyuz 2.1.a Fregat	Baikonour	
23 Oct	Progress M-58	Soyuz U	Baikonour	
26 Oct	XM-4 Blues	Zenit 3SL	*Odyssey* platform	
8 Nov	Arabsat 4B	Proton M	Baikonour	
12 Dec	Measat 3	Proton M	Baikonour	
19 Dec	SAR Lupe 1	Cosmos 3M	Plesetsk	
24 Dec	Meridian	Soyuz 2.1.a Fregat	Plesetsk	
25 Dec	Cosmos 2424-6	Proton K/block D	Baikonour	GLONASS
27 Dec	COROT	Soyuz 2.1.b	Baikonour	

Index